W0017811

Praise for *Riding to Arms*

"*Riding to Arms* carefully navigates the complex, sometimes complementary and sometimes contradictory, relationship between high school equitation and practical military riding, highlighting and quoting from the works of all the great masters of horsemanship. Taking the reader through the development of the art of the *manège* (school) style of riding and the technical aspects of military horsemanship, this book should be on the library shelf of all serious students of riding technique and cavalry."—Louis A. DiMarco, LTC (Ret.) U.S. Army, author of *War Horse: A History of the Military Horse and Rider*

"This wide-ranging and well-written book offers new insights into the emergence of effective light cavalry in western armies from the eighteenth to the early twentieth centuries. Authored by a lover of horses and riding, it draws on intellectual, practical, and cultural writings to highlight an increasing understanding of the horse as a living animal and as the key to the mounted warfare and mobility of that era."—Stephen Badsey, author of *Doctrine and Reform in the British Cavalry 1880–1918*

"Charles Caramello surveys four centuries of equestrian literature from early works on *manège* riding to works on conditioning horse and soldier for modern cavalry warfare. A historiography of ideas about horsemanship, where proponents of different schools often sparred over training philosophies, *Riding to Arms* traces the literature from the *manège* treatises of the sixteenth through eighteenth centuries to the cavalry manuals of the long nineteenth century, culminating in the Great War."—Elizabeth Tobey, editor and cotranslator of *Federico Grisone's* The Rules of Riding: *An Edited Translation of the First Renaissance Treatise on Classical Horsemanship*

Riding to Arms

Riding to Arms

*A History of Horsemanship and
Mounted Warfare*

Charles Caramello

UNIVERSITY PRESS OF KENTUCKY

Copyright © 2022 by The University Press of Kentucky

Scholarly publisher for the Commonwealth,
serving Bellarmine University, Berea College, Centre
College of Kentucky, Eastern Kentucky University,
The Filson Historical Society, Georgetown College,
Kentucky Historical Society, Kentucky State University,
Morehead State University, Murray State University,
Northern Kentucky University, Transylvania University,
University of Kentucky, University of Louisville, and
Western Kentucky University.

All rights reserved.

Editorial and Sales Offices: The University Press of Kentucky
663 South Limestone Street, Lexington, Kentucky 40508-4008
www.kentuckypress.com

Illustrations credit: Unless otherwise noted, all images are provided by and used with the permission of Alamy Stock Photo.

Library of Congress Cataloging-in-Publication Data

Names: Caramello, Charles, author.
Title: Riding to arms : a history of horsemanship and mounted warfare /
    Charles Caramello.
Other titles: History of horsemanship and mounted warfare
Description: Lexington, Kentucky : University Press of Kentucky, [2021] |
    Series: Horses in history | Includes bibliographical references and index.
Identifiers: LCCN 2021042888 | ISBN 9780813182308 (hardcover) |
    ISBN 9780813182315 (pdf) | ISBN 9780813182322 (epub)
Subjects: LCSH: Cavalry--History. | Horsemanship—History. | Cavalry—
    Bibliography. | War horses--History. | Horses in literature.
Classification: LCC UE145 .C28 2021 | DDC 357/.1—dc23

This book is printed on acid-free paper meeting
the requirements of the American National Standard
for Permanence in Paper for Printed Library Materials.

Manufactured in the United States of America

 Member of the Association of
University Presses

For Anne and Dagmar . . .

Contents

Illustrations follow page 134

Preface

Unhorsed in battle, Richard III cries, "A horse, a horse, my kingdom for a horse!" Though long since flattened into cliché, Shakespeare's once fresh conceit called on a nascent early modern conception of the horse's critical role in warfare. Some forty years before *Richard III* (1593), Federico Grisone had based the first modern treatise on horsemanship, *Gli ordini di cavalcare* (1550), translated as *The Rules of Riding*, on the premise that the systematic training of horses, or dressage, and that of riders, or equitation, not only determined success in mounted warfare, but also, as a consequence, directly affected the prestige of rulers and the power of states. Grisone's premise would hold in Europe and then North America, with various inflections, for half a millennium.

Prior to Grisone, in the chivalric age of the fourteenth and fifteenth centuries, as Friedrich Wilhelm Bismarck pointed out in 1818, "The cavalry, consisting of the nobility, constituted . . . the flower of the army, and was the support of princes and of their kingdoms." Once Grisone had refined and codified the preparation of horse and rider in the mid-sixteenth century, enabling cavalry to provide such support more effectively, formal horsemanship and its martial application thenceforth evolved in tandem. Massari Malatesta's *Compendium of the Heroic Art of Chivalry* (1599), for example, synthesized the principles of Grisone and other early theorists and added "an overview of the use of the horse for military purposes," in Giovanni Tomassini's words, an overview that "underlines the importance of the horse's agility and obedience" in single mounted combat. Whether or not Malatesta was the first horseman to emphasize the necessity of highly schooled (or "educated") horses for warfare, he was far from the last.[1]

Mounted military bodies in Europe following the sixteenth century evolved from the primarily *heavy cavalry* of the seventeenth century, through the predominant *light cavalry* of the eighteenth and nineteenth centuries, to the resurgence of *dragoons*, who fought mounted and unmounted, in the late nineteenth and early twentieth centuries.[2] Concurrently and reciprocally, the discipline of formal horsemanship in Europe evolved from its initial emphasis on both art and utility, through a lengthy stage of increased emphasis on the artistic and decreased emphasis on the utilitarian, to a final rebalancing of emphases for optimal effectiveness in warfare. Maintaining that effectiveness, in turn, led to the pivotal blending of formal (or *manège*) and cross-country training of horse and rider that dominated military horsemanship in the nineteenth century. As a consequence, in the words of the twentieth-century horse soldier and eminent horseman, Vladimir Littauer, "The old incentive to 'ride the way princes ride' was replaced by the new one, 'To ride the way the cavalry rides.'"[3]

The cavalry "way of riding," however, was neither a static nor a simple concept. Early theorists of dressage and equitation may have intended formal *haute école* (or "high school") training of horses to serve military purposes and to enhance the aesthetic refinement and social prestige of nobility, but, as warfare evolved, theorists debated the practicality and feasibility of such training for cavalry. Colonel Paul Rodzianko's distinction in *Modern Horsemanship* (1936) between high school training that "requires artificial movements from the horse" and *advanced training* "based on natural movements" helps clarify the terms of that debate. Introduced "many years ago for military horses who had to be perfectly controlled as they moved in large numbers," advanced training not only eclipsed high school, but also merged with cross-country training to produce military horses who would be tireless, fast, and agile as they moved in smaller numbers in detached units.[4]

Throughout the developments limned above and detailed in the following chapters, European militaries did not train, maintain, and field only saddle horses for cavalry use. Over the modern era, far more warhorses served as workhorses, primarily of draft breeding or crossbreeding, that provided motive and tractive power for all three military arms— infantry, artillery, and cavalry. In even the mechanized Great War of 1914–1918, equines remained the primary source of energy. Despite the introduction of motorcycles and automobiles, trucks and tractors, and eventually armed and armored tanks, "no one would have seriously suggested," as David Kenyon has observed of the conflict, "that the horse might be removed from the military scene altogether. . . . The presence of horses was taken for granted."[5] As in preceding wars, a minority of horses in the Great War served as cavalry chargers and troop mounts, while the majority transported personnel, supplies, and munitions to battle; hauled guns and their accoutrements in battle; and carried the wounded, maimed, and dead from battle.

European and North American militaries deployed horses, or "remounts," in untold numbers throughout a "long nineteenth century" that included Napoleonic Wars; British Crimean and Colonial Wars; American Civil and Indian Wars; Franco-Prussian and Boer Wars; and, finally, Great War.[6] The potential scale and scope of the Great War, though underestimated at first, still required a vast initial mobilization of equines, and the modern firepower and wholesale slaughter of the war itself demanded a steady flow of replacements. Millions of equines—horses and mules—served and died in the Great War, torn apart by shells or drowned in mud, done in by poison gas or by sheer exhaustion: the "wastage" of horses, as it was called officially, numbered in the millions. In the postwar years and decades, for better or worse, militaries inexorably turned from saddle and draft horses to mechanized vehicles. With respect to mounted cavalry, Germany and Russia fielded large forces in the Second World War, and the British Empire and the United States residual forces, but the age of *horse* cavalry in the anglophone nations, though not of cavalry itself, essentially ended in the 1930s. Sooner or later, all national cavalries dehorsed. "What is amazing," Louis A. DiMarco writes, "is not that the horse disappeared from military history, but that it lasted so long and then disappeared so swiftly."[7]

Riding to Arms: A History of Horsemanship and Mounted Warfare comprises five chapters—studies in the intertwined histories of formal horsemanship, particularly *manège* dressage and equitation (high school and advanced) in the sixteenth through nineteenth centuries, and mounted warfare, particularly light cavalry in the eighteenth and nineteenth centuries, together with their last major convergence in the Great War of 1914–1918 in the early twentieth century. While these chapters can stand individually, they proffer collectively a detailed picture of a complex historical evolution—a study of "the horse in history," rather than a history of the horse. They speak, primarily, to academic scholars and students of equine or military history, and, secondarily, to general horse enthusiasts and cavalry buffs. They aim to provide equine historians with background in the military employment of the horse in modern war and historians of mounted warfare with background in the modern theories of horsemanship behind cavalry training.

A book about books, its most distinctive and, arguably, its unique contribution to the conversation, *Riding to Arms* views equine and military history through seminal works written by theorists and practitioners during that history—books that shaped the aspects of that history relevant to our concerns. Meant to be illustrative rather than comprehensive, the selection of books discussed includes established classics on horsemanship and cavalry, as well as works less influential or renowned but amply deserving of reclamation and revival. Academic specialists in either equine or cavalry history will encounter many books already known to them, though perhaps in unfamiliar contexts that cast them in a different light. General readers, particularly those in the horse community, likewise will encounter books known to them, but, I believe, many more books rarely read if even remembered—works that collectively enable us to recover the origin, evolution, and cultural importance of modern equestrian sports.

The selection of books totals nearly a hundred titles published in Europe and North America between 1550 and 1950, with the preponderance published in the eighteenth and nineteenth centuries. Since I treat these works on the subjects of horsemanship and cavalry as contemporary historical documents, and treat the horsemen and horse soldiers who wrote them primarily as thinkers and writers, *Riding to Arms* constitutes, in effect, a study in the history of ideas about horsemanship and cavalry more than a study in equine or military history per se—a study that includes both history and historiography, biography and bibliography. Not all ideas, of course, find their expression in print or in works of nonfiction—in this case, principally treatises and histories—so, in later chapters, *Riding to Arms* extends its source materials to include contemporary visual representations, such as recruitment posters and wartime photographs and paintings, as well as examples of contemporary memoirs and works of fiction.

Leaving aside the countless books on military strategy, tactics, battles, and arms published over four centuries, printed books on horses, horsemanship, and horse management, including their military applications, emerged in numbers over the sixteenth and seventeenth centuries and then proliferated in the eighteenth century. Scores appeared in Great Britain alone in that century—English translations of Continental works, as well as works originally written in English. The sheer number that treated equine breeding, management, and care testifies to the universal importance of the horse in agriculture, industry, transportation, sport, and warfare. Reflecting the rationalistic, scientific, and encyclopedic eighteenth-century zeitgeist, they claimed reason and empirical evidence as their bases, and "useful knowledge" as their value. Comparable in number, though probably higher in ratio of translations to works written in English, treatises on dressage and equitation also claimed experiential validity and utilitarian value, but with an accent on aesthetics—as suggested by the frequency of the word "art" in their titles. Numbers of both general types of book exploded in the nineteenth century.

The bibliography of primary works related to horses is huge. F. H. Huth's *Works on Horses and Equitation: A Bibliographical Record of Hippology* (1887), for example, listed works, often multiple works, by over 1,800 authors; Anne Grimshaw's *The Horse: A Bibliography of British Books 1851–1976* (1982), contains over 3,000 entries; and A. Henry Higginson's *British and American Sporting Authors* (1949) treats 346 major writers, spanning 500 years, on foxhunting alone. John B. Podeschi's *Books on the Horse and Horsemanship, 1400–1941* (1981), catalogues 513 books and periodicals in the library of Paul Mellon, with a focus "on the Thoroughbred horse, its antecedents, anatomy, physiology, breeding, training, and employment on the turf and in the field"; and Koert van der Horst's *Horseman as Bookman: The Library of Johan Dejager* (2014) provides descriptive analyses of 435 books and manuscripts in a private collection that "contains nearly all of the great horse books published [in Europe] in the first two and a half centuries after the invention of printing."[8]

Since horses played such a definitive role in warfare and statecraft from the sixteenth through the nineteenth centuries, most of the Continental, British, and North American books on horses, horsemanship, and horse management published in that period discussed mounted warfare to greater or lesser degrees of breadth and depth—including not only Grisone's foundational *Ordini*, but also the first printed work on horsemanship in English, an adaptation and translation of Grisone, Thomas Blundeville's

The Arte of Ryding and Breakinge Greate Horses (1560). Legion among these books were the copious treatises and manuals focused on cavalry, particularly light cavalry, or "light-horse," as it evolved over the eighteenth century—works whose numbers grew exponentially in the nineteenth century and continued into the early twentieth century. Though typically centered on the subjects of strategy and tactics, cavalry treatises and, especially, manuals, also addressed, sometimes predominantly, military horsemanship, or dressage and equitation, and "horsemastership," or horse management and care.[9]

Though smaller, the bibliography of secondary works on horses and mounted warfare is significant. Recent books on the warhorse, for example, range from compact one-volume overviews, such as J. M. Brereton's *The Horse in War* (1976) or Louis A. DiMarco's *War Horse: A History of the Military Horse and Rider* (2008), to expansive multivolume series, such as Ann Hyland's sequence of four books on *The Warhorse* from the medieval period to the "beginning of the second millennium" (1996–2010). Studies of modern cavalry, particularly British cavalry, are manifold, such as the Marquess of Anglesey's majestic eight-volume *A History of the British Cavalry, 1816–1919* (1973–1997); and books on British cavalry in the Great War alone, finally, include not only numerous official and regimental histories and academic military histories, but also important revisionist histories, such as Stephen Badsey's *Doctrine and Reform in the British Cavalry, 1880–1918* (2008); David Kenyon's *Horsemen in No Man's Land: British Cavalry and Trench Warfare, 1914–1918* (2011); and Graham Winton's *"Theirs Not to Reason Why": Horsing the British Army, 1875–1925* (2013).

Riding to Arms draws its main theme from the evolution sketched earlier. In brief, the value of the horse in warfare depended not only upon proper breeding and management, but also, with respect to cavalry, upon proper training of the saddle horse and its rider. Heavy cavalry found its appropriate training in classical dressage and equitation, but light cavalry demanded the new "outdoor" horsemanship that melded *manège* and cross-country training. The value of the horse in warfare, additionally and concurrently, extended to the needs of all three arms for motive and tractive power. While the training of the draft horse differed from that of the saddle horse, the scale, scope, and conditions of mass industrial warfare, as in the Great War, required and wasted both draft and saddle horses in vast numbers.

Following that war, both horse cavalry and horsepower became increasingly anachronistic, and representations of them elegiac, but cavalry horses and horsemanship found a new life in military and civilian equestrian competition. A straight line, in a word, runs from Grisone through modern cavalry to contemporary horsemanship.

Riding to Arms advances that overall theme through three distinct but closely related arguments: one, formal horsemanship evolved together with mounted warfare, and with clear reciprocal influence, for over four centuries; two, military horsemanship (and horse management) accompanied the gradual but decisive transformation in cavalry tactics, over the long nineteenth century, from *shock action*, or battlefield charges in close formation with steel weapons, to *fire action*, or reconnaissance duties in detached patrols with firearms; and three, subsequent to that transformation, the Great War introduced mass, mechanized warfare that wasted horses in the millions; presaged the end of horse cavalry on a grand scale; and not only prompted postwar nostalgia for cavalry exploits and for field sports, particularly foxhunting, associated with cavalry, but also issued in postwar cultivation of civilian horsemanship and equestrian competition.

The following chapters hew to a roughly chronological order in developing those three arguments and in organizing this study's primary source materials—writings from the era:

Chapter 1, "Ryding and Breakinge," traces horsemanship and its military application from the work of Federico Grisone, the founder of modern horsemanship in the sixteenth century, through classical seventeenth- and eighteenth-century treatises by the giants Antoine de Pluvinel, William Cavendish (1st Duke of Newcastle), and François Robichon de la Guérinière, to Richard Berenger's summary *The History and Art of Horsemanship* (1771). It contends that the "art" of horsemanship, originally developed for warfare, as well as for aesthetic value and social status, migrated over this period from the former purpose to the latter.

Chapter 2, "*Manège* to Field," focuses on military dressage and equitation from the late eighteenth century to the early twentieth century with emphasis on the cavalry's merging of French *manège* riding and training with English cross-country hunt riding, and on the mutual influences and public rivalries among cavalry officers and theorists of horsemanship such as François Baucher, Le Comte d'Aure, Gustav Steinbrecht, and others. It contends that modern cavalry theory and practice influenced not only late eighteenth-century dressage and equitation, but also the "new methods" that emerged in the mid-nineteenth century and that returned the training of horse and rider from the aesthetic to the practical.

Shifting from horsemen and military application to horse soldiers and cavalry theory and practice, chapter 3, "Light-Horse, Dragoons, and Others," the longest chapter, traces the rise and fall of horse cavalry over the long nineteenth century, mainly in Great Britain and the United States, using period treatises and histories on military horsemanship and cavalry tactics, and ending with Great War postmortems on the future of horse cavalry in mechanized warfare. It focuses mainly, though not exclusively, on an ongoing debate between traditionalists advocating mounted fighting with sabers and reformists advocating dismounted fighting with carbines or rifles, a debate culminating in the *"arme blanche* controversy" of the pre–Great War years.

Chapter 4, "Remounts and Wastage," employs sources in equine and military history ranging from contemporary government reports and proposals, to popular and fine art and photography, to fictional and nonfictional literature. Unlike the first three chapters, this chapter focuses on horses rather than on horsemanship. Treating horses both as war matériel and as cultural symbol, it discusses the breeding, acquisition, transportation, deployment, and massive wastage of horses (loss by death or other causes) in cavalry and artillery in the Great War, as well as the representation of warhorses in Great War art and literature.

Chapter 5, "Hunting in the Trenches," finally, compares and contrasts British foxhunting with trench warfare mainly through two semiautobiographical works of fiction by the Great War poet and memoirist Siegfried Sassoon: *Memoirs of a Fox-Hunting Man* (1928) and *Memoirs of an Infantry Officer* (1932). The chapter contends, one, that Sassoon echoed the many early and contemporary British military writers who advised cavalry officers, whether in rank or potentially in rank, to foxhunt in order to enhance both field skills and leadership qualities; and two, that Sassoon, in the end, ambiguously affirmed while reproving the analogy and the advice derived from it.

It goes without saying that horsemanship, as a robust scientific and artistic discipline, has continued unabated throughout the twentieth century and into the present moment, and that cavalry, as an active military arm, likewise has continued up to and into the present through its armored, airborne, and now robotic incarnations. As compelling as those recent histories may be, *Riding to Arms* logically ends in the post–Great War decades of the 1920s and 1930s—in the interwar years when horsemanship began its shift from a primarily military to a primarily civilian enterprise, and when horse cavalry, for most practical purposes, became armored cavalry. Put differently, once horsemanship effectively ceased to have military application, and once cavalry effectively ceased to be horse cavalry, they left the purview of this study.

Since *Riding to Arms* aims, in part, to reclaim primary works, it uses their language liberally, but avoids large swatches of quotation by weaving their words together with my own, employing ellipses when eliding words and brackets when inserting paraphrases. To convey original textural flavor, I retain the antique spelling, capitalization, and punctuation of early texts and the British spelling conventions of British publications.[10] Sources for quotations are given in the notes and are keyed to the bibliography of primary and secondary works; the notes themselves are variously documentary, explanatory, and supplementary; and sources and other ancillary information pertaining to a given paragraph generally are grouped within one note. Finally, the authors and actors in this history were virtually all men, and, from our vantage, their language is regressive: soldiers, as opposed to officers, for example, are "men," and horses, at least pronominally, masculine. For consistency, and I hope not controversially, I have followed those practices.

Some ideas and passages in this book were tested elsewhere. The discussion of Pembroke and Tyndale in chapter 3 derives from the introduction and notes to my edition *Eighteenth Century Military Equitation* (Xenophon, 2018). Chapter 4 adapts some material from two articles: "'They Had No Choice': Images of Equines in the Great War," *Anglistik: International Journal of English Studies* 29, no. 2 (2018); and "Hemingway's Horses," *Denver Quarterly* 47, no. 1 (2012). Brief extracts from chapter 5 have appeared on the website *Foxhunting Life*; and readings of several primary texts scattered throughout the book appeared in different form on the website *Horse Talk*, published in New Zealand, and in *Drawing Covert*, a forum on the website of the National Sporting Library & Museum. I want to thank all editors concerned for their confidence and support and for their permission to reprint material.

Generous institutional support made this book possible. I am grateful to the National Sporting Library & Museum in Middleburg, Virginia, at once a public library and museum and a research library with extensive holdings in rare books, periodicals, and archival materials. NSLM has supported my work with invaluable resources of many kinds—the expertise, enthusiasm, and friendship of its staff not the least of them. It is my privilege to conduct research there as a John H. Daniels Fellow, and I thank the board of directors and the executive director for honoring me with that continuing appointment. I also thank the University of Maryland, my professional home for four decades, for supporting this work with a Research and Scholarship Award, research assignments, and sabbatical leave. Finally, I thank the Press Committee and Editorial Board of the University Press of Kentucky for their enthusiasm in accepting this project for publication and the UPK editorial and production staffs for their professionalism in bringing it to completion.

Many individuals, finally, afforded invaluable assistance in researching and writing this book. Vin Carretta and John Yurechko provided expert information and advice on eighteenth-century culture and warfare; Jack Bryer and the late Stan Plumly offered seasoned editorial counsel on both form and content; and Jim Wofford shared his deep, encyclopedic, and insightful knowledge of horsemanship and cavalry. My daughter, Dagmar, introduced me to horses, and my wife, Anne, encouraged this work on them. Many friends in the Maryland and Virginia horse communities shared their practical knowledge of dressage and equitation, as have "some horses" (in Thomas McGuane's nice phrase). My deepest gratitude goes out to all.

1

Ryding and Breakinge

What Sets-off a King more, than to be upon a beautiful and ready
horse, at the Head of his Army.

Sir William Hope, *The Complete Horseman* (1696)
Translated from Jacques de Solleysel, *Le parfait Mareschal* (1664)

Classical horsemanship was revived in Italy in the sixteenth century and,
through a succession of weighty and influential riding masters, migrated
to France in the seventeenth and eighteenth centuries: the dominant
bloodline comprised the progenitor Federico Grisone and a handful of his
fellow masters in the late sixteenth-century Neapolitan school; followed
by their younger contemporary, the great horseman Giambattista Pignatelli, and his two brilliant early seventeenth-century French students,
Salomon de la Broue and Antoine de Pluvinel. From that point, Giovanni
Tomassini writes in *The Italian Tradition of Equestrian Art* (2014), "Italy
quickly [lost] the hegemony that it hitherto had in the equestrian field,"
and eventually, in the mid-eighteenth century, Vladimir Littauer adds
in *Horseman's Progress* (1962), France arrived at "the acme of early High
School" in the person of François Robichon de la Guérinière, who not
only brought classical dressage to its pinnacle, but also presaged the emergence of modern dressage.[1]

Master horsemen throughout those three centuries (master horsewomen were not yet on the recorded scene) formed not only a succession
of equestrian academicians and their pupils, but also a close, elite, and cosmopolitan society: generally wellborn and well connected, cultivated and
polyglot, they circulated freely, as peripatetic as their ideas, through "lucra-
tive, prestigious positions in the major European courts."[2] Masters and
disciples influenced one another both vertically through their succession

and horizontally across their community in ways direct and indirect, acknowledged and unacknowledged, obvious and subtle. Their theories of dressage and equitation collectively informed the training of the horse and its rider for purposes of war, art, and social prestige. Since early horsemanship, moreover, was embedded in its zeitgeist, the discipline's masters developed equestrian principles and academies that not only reflected, but also affected, multiple spheres of thought and action.

Precursors

Xenophon of Athens (c. 431–354 BC), military commander, philosopher, and historian, wrote the foundational *Art of Horsemanship* (also known as *On Horsemanship*) circa 362 BC. "The oldest extant work on the subject in any language, and the only one which has come down to us in either Greek or Latin," in the words of its late nineteenth-century English translator, Morris H. Morgan, *The Art of Horsemanship* was among the classical texts from many intellectual disciplines recovered and revived by Renaissance thinkers.[3] Though not cited by Federico Grisone in his early and seminal Renaissance treatise, *Gli ordini di cavalcare* (*The Rules of Riding*) (1550), Xenophon's *Art of Horsemanship* influenced not only Renaissance equestrian theory and practice, but also the vast body of European writing on dressage and equitation written from the Renaissance to the present.

Xenophon's influence and longevity derive from two sources. First, as Littauer contended, despite "the briefness of Xenophon's writing and the primitiveness of his techniques . . . many of his suggestions are immortal simply because they are based on an understanding of the horse's psychology, which has probably changed little since his days." While Xenophon did not offer a systematic method for the training of the horse and, moreover, *could not* offer a scientific treatise on the basis for such training, his work "can be considered as the first attempt to *reason* about riding that has come down to us." Second, as W. Sidney Felton pointed out in *Masters of Equitation* (also 1962), "Xenophon's approach to schooling," as both a reflection and an agent of the enlightened culture of the Greek golden age, was itself "gentle and enlightened." As Morgan had put it, "no modern humanitarian was ever more earnest in urging over and over again the principle of treating horses with kindness."[4]

Machines of battle

Xenophon intended his text to be, above all, practical—a work of applied theory. Neither geography nor overall military strategy predisposed ancient Greece to practicing mounted warfare or to developing advanced cavalry tactics, so despite the status of Athenian cavalry as "a *corps d'elite*," Morgan argued, "the Greeks never accomplished the revolution in military art which gave cavalry a decisive role in action." As Littauer would put it, the "main military strength [of classical Greece] consisted in its navy and in its foot soldiers." Nevertheless, as Morgan went on, the ancient Greeks bred and reared horses primarily as warhorses, as "machines of battle," and devoted much thought and many resources to their development, even "the pomps and processions on festive days . . . so contrived as to be part of the horse's training for war." As a cavalry general writing in that context, Xenophon naturally devoted *The Art of Horsemanship* to the practice of preparing horse and rider for battle.[5]

Morgan's reference to horses as "machines," however, misses the mark. Though Xenophon does not invest the warhorse with the anthropomorphic status of fellow warrior, as much later cavalry writers would do, he works from the premise that horses are sentient beings that, with proper selection, training, and riding, willingly serve human warriors to mutual and reciprocal benefit. The proper warhorse, in his view, should be "sound-footed, gentle, sufficiently fleet, ready and able to undergo fatigue, and, first and foremost, obedient." It also should have a "fierce," "magnificent," and "striking" appearance, for purposes of intimidation, and should display appropriate "mettle." Since "mettle is to a horse what temper is to a man," however, a "too high-mettled horse," like a very hot-tempered man, is not suitable for battle. Though some of these traits are innate (hence the importance of proper breeding and selection), most are cultivated (hence the greater importance of proper training and riding).[6]

Xenophon's text, as its title suggests, focuses on horsemanship (and thus horsemen) far more than on horses. He excuses the proper cavalryman from breaking his own horses, but not from teaching his groom "the proper way to treat the horse," nor from teaching "his own horse" how to ride cross-country. The cavalryman, he proposes, should be able to mount quickly from both on- and off-side, should be able to ride "as though he were standing upright with his legs apart" (and not in a chair seat), and should be able to ride with collection into *voltes* and reach speed quickly after turning. Most important, the cavalryman must be as dispassionate as his horse is obedient: "The one great precept and practice in using a horse is this,—never deal with him when you are in a fit of passion." In essence, equine obedience presupposes human patience and kindness. *That* is Xenophon's principal lesson and his legacy.[7]

Xenophon notes in his closing line that he wrote *The Art of Horsemanship* "for the private [trooper]," having "set forth already in a different work" what he calls "the knowledge and practice necessary for the commander of cavalry."[8] That work, dating to the same period as *The Art of Horsemanship* but obviously preceding it, is the treatise called *The Cavalry General*. As its title suggests, this work does not instruct cavalrymen in horsemanship, but rather instructs cavalry commanders in leadership: as soldiers are to horses in the one, so commanders are to soldiers in the other. Though much shorter than its successor volume, and far less renowned, *The Cavalry General* retains timeliness for anyone interested in mounted warfare through its sound reading of the soldier's psychology, its reasoned approach to military command, and its enlightened view of leadership.

With appropriate devoutness, Xenophon immediately informs his intended reader, the military commander, that "your first duty is to offer sacrifice, petitioning the gods," and only then, after obtaining their goodwill, to "proceed to mount your troopers" and, following that, to equip them with horses that are "well fed and in condition . . . and are tractable [since] a horse that will not obey is only fighting for the enemy and not his friends"—a point that he would repeat, with others, in *The Art of Horsemanship*.[9] Once troopers have learned basic equestrian skills, a commander can train them in *evolutions*, or movements of troops, that will serve them "in processions," in "maneuvers," in "real battle," and on the march—in short, in all mounted actions requiring the mass orderliness that only disciplined and skilled individual troopers can ensure. As Xenophon notes, a general can and should delegate this training of horses and men to subordinate officers.

Some duties, however, "devolve upon the general of cavalry himself in person," such as offering sacrifices to the gods on behalf of the cavalry; setting a "personal example" for officers and troopers; possessing technical knowledge and sound judgment; and, especially, thinking both strategically and tactically. A proper general, for example, must know technical niceties, such as "within what distance a horse can overhaul a man on foot, or the interval necessary to enable a slower horse to escape one more fleet." More important, he must possess the judgment to appreciate that playing "into the enemy's hands may more fitly be described as treason to one's fellow-combatants than true manliness"; or that "if you attack with a prospect of superiority, do not grudge employing all the power at your command [since] excess of victory never yet caused any conqueror one pang of remorse." A cavalry general's quality of mind and strength of character, in short, carry as much weight in achieving victory, if not more weight, than do the quality and strength of his horses and men.

The salient feature of that mind and character is inventiveness, or creativity.[10] The general must be "a man of invention, ready of device to turn all circumstances to account"; a thief and trickster, who knows "not only how to steal an enemy's position, but by a master stroke of cunning to spirit his own cavalry away, and, when least expected, deliver his attack"; an illusionist, who can make a small force look large, or a large one small, and who can make his force seem near when far, or far when near; and a master of psychology, who, in a weak position, can convince the enemy, by feigning intimidation, *not* to attack, or, in a strong position, can entice the enemy, by dissembling weakness, *to* attack. In a word, though "inventiveness is a personal matter, beyond all formulas—the true general must be able to take in, deceive, decoy, delude his adversary at every turn, as the particular occasion demands. In fact, there is no instrument of war more cunning than chicanery."

Xenophon portrays his ideal cavalry general, in sum, as a sagacious, judicious, courageous, charismatic, and cunning leader who appoints capable officers as deputies; who recruits troopers that not only will "practice of their own accord the art of horsemanship," but also "will dash forward and charge an enemy . . . with an eager spirit and unfailing courage"; and who ensures that they purchase quality horses and capably oversee their training. A commander who fails to discharge those duties invites catastrophe, for when ill-prepared men riding poor horses face a superior force, "the worse mounted will be captured," the unskilled "will be thrown," and the others "will be cut off owing to mere difficulties of ground." Once engaged, moreover, ill-prepared men and mounts will produce "collisions" and "kicks through mutual entanglement" that inflict as much damage as does the enemy. Perhaps most important, as Xenophon admonishes in closing, a general must *do* as well as *think,* and must *do* with determination, for "without pains applied to bring the matter to perfection, the best theories in the world . . . will be fruitless."

Riding and living

Needless to say, neither horsemanship nor the employment of horses in warfare disappeared or remained stagnant in Europe in the roughly seventeen centuries between Xenophon and Grisone. Advances included, for example, introduction of "the saddle, the stirrups and the horseshoe," and "the use of the legs to turn a horse, not merely to make him increase his pace." From the Roman imperial age through the European "dark ages," educators of horses and riders and strategists for their military use continued to teach and to fight, clearly with intelligence and innovation, transmitting what they learned orally and surely, in some cases, recording it in writing. We can reconstruct much of that evolution through related historical documents and artifacts, but these horsemen lived in a manuscript culture and, sadly because not inevitably, their manuscripts on the whole did not survive them.[11]

We know, for example, that heavily armored knights in late medieval and early Renaissance Europe fought (and jousted) on Great Horses, employing rudimentary equitation.[12] We also know that these practices in horsemanship were embedded in, and were signal contributors to, a chivalric code of ethics and manners that reflected, informed, and largely defined the social and political culture. A professional way of life for knights, proper horsemanship served as a metaphor for a proper way of life for all cultivated persons. One manuscript that did survive provides us with a detailed account of chivalric equestrian practices and of their metaphoric value in one country: *The Art of Riding on Every Saddle,* written in the 1430s by Dom Duarte (1391–1438), king of Portugal.[13]

A didactic treatise about riding and living, *The Art of Riding* focuses on knights, or the riders, and not on Great Horses, or the ridden. In manifest content, it offers a study in equitation rather than dressage. While a knight with "natural talent or vocation" has an advantage, that is, all knights can (and must) become accomplished by "having good horses," by acquiring knowledge through the example of expert horsemen, by "continuously practicing the art," and by gaining courage and tranquility while eliminating fear.[14] In latent content, it offers a parallel manual for exemplary behavior and moral growth. While persons of noble birth may enjoy varying degrees of privilege (in effect, their "natural talent"), they all can become persons of virtue and stature by emulating knights and, metaphorically, their equestrian skills, and by adhering to knightly, or chivalric, protocols for behavior.

Dom Duarte focuses on equitation for tournaments (competitions between groups of knights) and jousts (competitions between individuals), as well as for combat. Equitation, in this context, included use of weapons while mounted, since skill in using weapons was "inseparable from skill in riding" in the early modern period, as Tomassini notes. Dom Duarte establishes the exigency for such skilled equitation both early in his text—"riding skills are one of the most valuable skills for warriors"—and also late—"it should not be forgotten that the arts that are the most important for combat, are the ones that we should—above all others—learn and master." Good horses are prerequisite—"do not forget that, above everything, you must have a good horse; without it, all the knowledge you have and the preparations you have done are of very small value"—but knowledge and practice are essential. Dom Duarte waffles on the value of reading about equitation, including his own book, but nowhere on the value of training both in equitation and in hunting and other analogous athletic endeavors.[15]

Dom Duarte seems neither overly prescriptive nor proscriptive regarding technique, allowing that different solutions can be equally effective and that "every man has his own preferences."[16] His principles for a proper knight's (or any proper rider's) character traits and equestrian goals, however, are close to universal. A knight should have "greatness of heart . . . the right level of self-esteem and self-confidence . . . daring and perseverance," and a willingness to "take chances."[17] A knight should be ruled by "reason," rather than feelings or nature (that is, sentiment or passion), in both general behavior and horsemanship. A knight should seek harmony, "to unite [himself] with the horse's body," and "should never forget that his main objective is—above everything—to stay firm and erect on the beast."[18] When Dom Duarte proposes, finally, that a knight should achieve outer "quietness" and inner "tranquility," he illustrates perhaps the key point underlying his text.[19]

Appearances mattered in the chivalric tradition, that is, as well as in the Renaissance humanism already established in Italy (if not Portugal) by the mid-fifteenth century. Dom Duarte's emphasis on the "elegance and handsomeness" of a proper leg, to cite just one example, reflects his belief that "all the feelings and sentiments we have in our hearts are known only to ourselves, whereas everybody knows and evaluates us only by our acts." This idea that a person can see only the appearance and acts of another person and therefore must infer character and intentions from them adumbrates the idea that appearance and character are mutually informing and reflective—the basic premise behind the Renaissance development of portraiture as a genre. Dom Duarte makes a number of explicit moral analogies in his text, the most salient, if most obvious, his drawing from "our objective of staying erect when mounted on the beast" the lesson that "we should also behave in life acting in such a way that we stay firmly mounted on it." It is his point about outward appearance and inward character, however, that links him closely to Grisone and the birth of modern equitation.[20]

Progenitors

If Xenophon and Dom Duarte were the precursors of modern horsemanship, its distant and near ancestors, then Grisone (dates unknown) and his Neapolitan contemporaries and successors were its progenitors, its "onlie begetters."[21] Grisone's *Gli ordini di cavalcare* (*The Rules of Riding*), the urtext of *manège* riding and modern dressage, was published by G. Suganappo, in Naples, in 1550. It was, as Tomassini notes, "the first equestrian treatise ever printed in modern times." Many editions of the *Ordini* in Italian, as well as translations into multiple languages, quickly followed, including Thomas Blundeville's loose English translation, *The Arte of Ryding and Breakinge Greate Horses*, in 1560.[22] The *Ordini*, in short, produced a powerful and immediate impact with long and broad historical resonance.

The *Ordini*'s importance derives from its originality both as a system of dressage and equitation and as a discourse on them. If we assume that the art of horsemanship not only was practiced between Xenophon and Grisone, but also was expounded in manuscript, particularly in the medieval period just prior to the Renaissance, then we might conclude that Grisone did not invent the art of *manège* equitation so much as codify and advance an already established art, as Charles Chenevix-Trench, among others, has noted in *A History of Horsemanship* (1970). In the end, that takes little from Grisone's achievement. As Elizabeth Tobey writes in the introduction to her translation of the *Ordini*—the first English translation since Blundeville's—"It was . . . Grisone who first explained in technical detail the training of the horse for the *manège* and the performance of its movements," though the practice of *manège* training predated "the *Ordini* by at least a half-century." In writing the first extended treatise on the systematic dressage of the horse, put differently, Grisone also was inventing the genre.[23]

The *Ordini*'s (and the emerging genre's) complexity and difficulty derive from the multiple purposes and values that Grisone assigned to equitation, dressage, and the educated riders and horses produced by them; to the proper methods for achieving those purposes and advancing those values; and to the proper discourse for systematically describing and teaching the methods and values. As Grisone notes, "I focused more on the correctness of the rules . . . than to words, so that everyone who reads will be taught more to ride than to talk." Like all writing on the subject, moreover, the *Ordini* must contend with the circular reciprocity of dressage and equitation: trained horses presuppose trained riders, and vice versa. Organized into four books, the *Ordini* unfolds a lengthy, detailed, often technical "systematic approach to the athletic development of the dressage horse," as David Guy notes in a preface, that anticipated by laying the foundation for specific elements in the teachings of major theorists to follow and also, eventually, for the basic elements of the twentieth-century Fédération Équestre Internationale (FEI) training scale as well.[24]

In the opening sentences of the *Ordini*, Grisone signals the dual purpose and value of dressage and equitation, and thus his system for them and treatise on them: *war* and *art*. With respect to war, advancements in weaponry necessitated new battlefield tactics, and these, in turn, necessitated new and more sophisticated training of horse and rider—training that would increase both the rider's control and the horse's agility. Grisone promotes both his overall system as providing that training efficiently and effectively, and its specific elements as producing advantages in large engagements and single combats—teaching the horse to turn on its haunches "so that his head will always be facing his enemy," to cite just one example, "is very useful in fighting man to man on horseback."[25] At the same time, though, he also seems to acknowledge that advanced *manège* training, with its sophisticated movements, or "airs," has limited if any martial application. Its purpose and value, then, also must lie—perhaps primarily must lie—elsewhere.

That "elsewhere" appears to be the realm of art. Grisone has proffered his written rules, he says in his dedication to Don Ippolito D'Este, "to redeem the art of riding from vile oppression," and he adds in his opening sentences proper, "In the military art, there is no discipline of greater beauty than that of horses, and none so adorned with handsome impressions. But it is also a useful art, endowed with great value. And this art is equally as difficult and worthy of praise as it requires one to employ *time* and *measure*" (my italics). Though Grisone tends to use "art" (or *arte*, in the original) in its multiple meanings, he borrows the concepts of time and measure, at the heart of his system, from Renaissance music and dance. He also depicts horse and rider off the battlefield as engaged in a performing art perhaps originally meant to imitate combat—"when you show the horse, you want to approximate the experience of the battlefield"—but eventually practiced as an aesthetic activity. Grisone is a Renaissance thinker, however, and not an aesthete: that activity has purpose and value that the *Ordini* explores at length.[26]

Art in the Renaissance, in other words, served ends ranging from the devotional to the representational, and, like other arts, equestrian art was *not* "art for art's sake," as writers such as Chenevix-Trench have suggested it was. Both Tobey and Tomassini have demonstrated in detail, for example, how mastery of equitation and dressage—as confirmed by educated horses—not only served nobles in "gaining social standing and respect at Renaissance courts" and became an essential "part of the training of young noblemen," but also led to "the birth of the equestrian academies," whose curricula included all the liberal arts, and to "the publication of the first treatises devoted entirely to the art of riding horses"—the *Ordini* being the prototype. The linkage of horsemanship and courtly advancement may invite skepticism or even cynicism, of course, but it also reflects deeply held moral and cultural values.[27]

The Renaissance, as its name signifies, witnessed the "rebirth" of Western culture through the revival of classical thought and art in all disciplines—including the revival of classical horsemanship as formulated by Xenophon. Renaissance aesthetic values, accordingly, include organic unity and harmony, proportion and moderation; and its aesthetic achievements include the use of "artificial" means to produce, or to simulate, "natural" results (much as the invention of perspective in painting enabled the representation and simulation of three dimensions on a two-dimensional surface).[28] Related Renaissance moral values include, among many others, nobility of character and behavior and consonance between them. Grisone reflects and advances all of these values in his prescriptions for the "one-ness" of horse and rider and for the invisibility of aids, as well as in his ideal of dressage as revealing and enhancing the "nobility" of the natural horse in the form of the educated horse.[29]

By way of analogy, Leonardo's sublime portrait, *Ginevra de' Benci* (1474/78), has an emblematic reverse side depicting a scroll with a Latin inscription meaning "beauty adorns virtue." The portrait and emblem together reflect the linkage of "female beauty and virtue . . . in Renaissance thought and art," based on "the Neoplatonic notion that physical beauty signified an inner beauty of spirit." Grisone reflects this notion when he writes: "And do not think that the horse, although he is well put together by nature, can work well on his own, without human aid and true teach-ing. It is necessary to awaken the parts of his body and the hidden virtues that are within him through means of the art of riding, and through good order and good discipline his goodness will become manifest to a greater or lesser degree." The horse in nature, in a word, can express inner virtue in the beauty of outer movement, but the virtue of the horse under saddle can express itself only through the compensatory artifice of dressage and equitation: "The rider," as Tobey puts it, "reveals his own virtue while also enabling his horse's own virtues to emerge." In this way, the aesthetics and ethics of riding become one.[30]

Not everyone would agree with such a positive reading of Grisone. Most commentary on the *Ordini* has characterized its end as the complete subjugation of the horse and its means as inhumane treatment of the horse manifested in cruel techniques and punishments.[31] Without question, that assessment has truth and its approbation has justice: Grisone recom-mends dealing with a horse who tries to lie down in water, for example, by holding his head underwater while beating him with a stick.[32] The same commentary, however, also tends to ignore or gloss over the frequency and specificity of Grisone's many attributions of a horse's bad behavior to the ignorance, and often cruelty, of its trainer; his repeated injunctions to respect a horse's capacity to understand and his reproofs against breaking a horse's spirit; and, finally, his admonishments to praise and reward a horse, as well as to correct and punish it.[33] Grisone's primary edict that "a rider needs to use punishment or aid and support in a timely manner and [with] restraint" carries aesthetic and ethical force: only those who do so "can well call themselves a most talented horseman" of virtue and valor.[34]

Horses of service, horses of pleasure

Though early modern British horsemen, like generations of their descendants, preferred hunting to schooling, "*manège* riding," as Tobey observes, "had been practiced at the English court since the early sixteenth century." Italian masters taught in England, British noblemen studied in Italy, Felton adds, and "the Neapolitan school," as a result, became known in England and excited "the interest of English horsemen." It was in this context that Thomas Blundeville (1522?–1606?), the British courtier and polymath, author or translator of ten books on diverse subjects, published his adaptation and translation of Grisone's *Ordini* as *The Arte of Ryding and Breakinge Great Horses* (1560). Blundeville's translation would enjoy "ten historical printings between 1560 and 1609" and, in Koert van der Horst's opinion, would influence "English horsemen for over a century"— an assessment that seems closer to the mark than Guy's assertion that the translation "did not have much of an influence on artistic riding in the English-speaking world due to lack of interest in dressage."[35]

Whether one considers Blundeville's *The Arte of Ryding* as an abridgment or as a reworking of the *Ordini*, Blundeville made clear in his dedicatory preface his rationale for translating the work "into our vulgare tounge." Grisone, in Blundeville's view, was "a farre better doer, then a writer," largely because he makes too "many repiticions of one thinge." As a consequence, Blundeville "thought how to bringe so good a matter . . . into a better forme . . . and thereby too make it the playner and also the briefer." With those goals in mind, Blundeville condensed Grisone's four books into three books on horsemanship, dressage, and vices and corrections. At the same time, he followed Grisone in appending to his text fifty plates representing "the shapes and figures, of many and diverse kyndes of Byttes, mete to serve divers mouthes," specifically, seventeen plates of "close byttes," twenty-five plates of "broken portes and upset mouthes," and eight plates of "wholle portes."[36]

Though a comparison of Blundeville's translation with Grisone's original text is not to our purpose, and has been capably done, in any case, by

Tobey, two of Blundeville's contributions warrant mention. First, Blundeville distinguishes in his preface between "horses of seruice" (that is, military service) and "horses of pleasure, called Styrers," on the basis of elements common to both and unique to each with respect to the horse (breeding, temperament, and so forth) and to its purpose (military service or pleasure) and training for its purpose. That binary distinction frames and informs Blundeville's signal point, second, that a service horse needs dressage, or training, but *not* dressage in advanced *manège* movements, or "airs." Superfluous, or even damaging, in a service horse, they are the proper pursuit of school riders and the proper goal for pleasure horses. On that basis, Blundeville advised in a subsequent edition of *The Arte of Ryding* that horses in the Queen's service should not be bred for airs, "but such as are onlie kept for pleasure, whereof it is sufficient to have in her highnesse stable two or three at the most."[37]

Soon after the first publication of *The Arte of Ryding*, Blundeville incorporated a "newly corrected and amended" version of it as the second tract in his comprehensive treatise, *The Foure Chiefest Offices belonging to Horsemanship* (1565/66). "After that I had put foorth the Art of Riding," he begins the latter's dedicatory epistle, "I sought to make the saide Booke more perfect, by adding thereunto three other Bookes, whereof the first shoulde trete of the breeding of Horses: the second of their diet, to preserue them long in health: and the third of their diseases, declaring therewithal the causes signes, & cures of the same. . . . For then it should bee a perfect worke, comprehending the foure chiefest Offices belonging to Horsemanshippe, that is to say: The office of the Breeder, of the Rider or Breaker, of the Keeper, and of the Ferrer."[38]

Blundeville's decision to produce a comprehensive work on horsemanship and farriery raises two points. First, Blundeville combined the genre of writing on the care of horses published prior to Grisone with the genre on the dressage of horses created by Grisone. His hybrid *The Foure Chiefest Offices* anticipated a robust tradition of such works, including notable seventeenth-century examples by Nicholas Morgan, Gervase Markham, and Jacques Solleysell. Second, Blundeville pressed much harder in *The Foure Chiefest Offices* than in *The Arte of Ryding* on the public value of horses and horsemanship—on service over pleasure. He thus advocates in the latter work, for example, "the necessarie breeding of Horses for seruice, whereof, this Realme [England] of all others at this instant hath greatest neede . . .

whereby this Realme shoulde bee of such force, as our enemies woulde alwaies be afraid to attempt any enterprise against vs."[39] For military service and national security, in short, total horsemanship trumps *manège* riding.

A very good year

In the preface to *The Arte of Ryding*, Blundeville advises readers to be thankful both to Grisone for having invented a system of horsemanship and to Blundeville himself for reordering and clarifying its presentation. He then adds: "And you shall haue very good cause also to be thankful unto my deare frende John Astley," who practiced Grisone's rules daily. "I sawe him without helpe of any other teacher, bring two of his horses . . . onto such perfection as I beleue few gentilmen in this realme haue the lyke." Blundeville's point is not exactly that his readers should thank Grisone for Astley's exceptional horsemanship, but rather, and obliquely, that they should thank Astley as the "deare frende" who brought Grisone's *Ordini* to Blundeville's attention and who then encouraged Blundeville to adapt it and translate it into English.[40]

John Astley (c. 1507–1596) was the son of "a Gentleman Pensioner, one of the ultra-elite royal bodyguards . . . charged with the breeding of war-horses and the cultivation of horsemanship in England."[41] To the manner born, Astley too became an accomplished horseman and an ensconced courtier appointed to multiple posts of importance following the accession of Queen Elizabeth to the throne in 1558, including Gentleman of the Privy Chamber and Master of Jewel House, Master of the Game in Elizabeth's Enfield Chase and Park, and steward and ranger of the manor of Enfield.[42] In addition, and more to the point, Astley published in 1584 the second book on horsemanship in English: *The Art of Riding*, "a breefe treatise," to cite its title page, "with a due interpretation of certeine places alledged out of *Xenophon*, and *Gryson* [Grisone], verie expert and excellent Horssemen." Brevity notwithstanding, *The Art of Riding* was signal in the emergence of systematic dressage and equitation in England.

At even this nascent moment of equestrian treatises, Astley could state his intention in his dedicatory "letter missiue" as neither giving new rules nor altering old ones, but rather as "interpreting, explaning or shewing the reasons of such rules" with reference to his own practice of them. The most important of those rules pertain to the mastery of "the true vse of the hand, wherein the chiefe substance of the whole Art of Riding standeth . . .

a thing not easie, but very hard to be understood"—a "thing" so important that Astley repeats variations of that phrase nine more times over the course of his treatise. Astley then names Xenophon and Grisone not only as skilled horsemen who mastered the true use of the hand, but also as the writers who have explained this "thing" with the most clarity and authority. Invoked throughout his text, they serve as exemplars whose precepts, or rules, "all those that are desirous to haue the true order and exercise of this Art" should follow.[43]

Explaining the "true use of the hand," Astley notes, requires that he first address the art of riding in full, for "how shal a man make another know what the true vse of the hand is in the Art of Riding, if first he dooth him not to vnderstand in general, what the verie Art it selfe in nature is?"[44] After defining the art in his opening chapters, Astley then can turn to his main topic, "the true use of the hand," or *contact*.[45] Through contact, as Astley derives its import from Grisone, "all the other qualities may be best brought to perfection, and the head and necke to great staiednes, the mouth to a sweete and perfect good staie, the which (to end withall) he counteth to be the verie foundation of the whole Art." While Grisone and Xenophon agree on the foundational nature of contact, however, they differ on the degree of control afforded by proper contact: contact serves "to mainteine [the horse] without giuing him anie libertie at all, as *Gryson* saith, though it seeme otherwise to *Xenophon*."[46]

Astley attributes specifically to Xenophon "the patterne that Art should imitate," namely, regaining, or simulating, through artifice the natural movement of the horse. The general Renaissance aesthetic ideal of art imitating nature, as Astley saw, carried ethical dimensions with special importance in the art of riding. Since "an Horse is the matter and subiect wherevpon this Art worketh, and is a creature sensible," the rider must embrace gentleness and patience and shun both violence and unruly passion as "contrarie to nature, which to content and please is the end of the whole art." Highly averse to impatience, intemperance, and force, and circumspect about rigid control, Astley defines the ultimate goal, *obedience*, as the horse's "readie willingnes to doo the will of him that dooth command." Astley's emphasis on the horse's willingness to obey (as opposed to the rider's strict control), his disgust at the use and, especially, abuse of harsh bits and other "violent waies," aligns Astley, in the end, more closely with Xenophon than with his primary source, Grisone.[47]

Astley introduces "this Art of Riding and Horsemanship," finally, as belonging "to the warre and feates of armes," but he soon after invokes Xenophon (and follows Grisone) on the point that educated riders, in their various roles, employ two types of horses: "the one, for the [military] seruice aforesaid, the other for pompe and triumph, the which we call stirring horses, the vse of which are verie profitable for this seruice, because they teach a man to sit surelie, comelie, and stronglie in his seate, which is no small helpe to him that must fight and serue on horssebacke: but of this last I meane not now to speake." Despite the convoluted language, Astley appears to be saying that two types of horse may have different purposes, but that *one* art of riding serves both purposes: that of war and that of pomp. That dual purpose, in the end, confirms the importance of the art.[48]

In addition to Astley's treatise, the year 1584 also saw the publication of Thomas Bedingfield's *The Art of Riding*, a translation, abridgment, and adaptation of Claudio Corte's important treatise, *Il cavallerizzo* (1562). Among the most influential of Grisone's Neapolitan contemporaries, Corte (1525–16??) resided in the English court in the 1570s and deeply impressed its leading horsemen.[49] Blundeville, for example, praised Corte as a "most excellent Rider," and "well learned, wise, courteous, and modest," and Astley invoked him (together with Grisone) when discussing how best to turn a horse.[50] Bedingfield (?–1613) most likely knew Corte both indirectly through the latter's writing and directly through his teaching, so he obliged when the Gentleman Pensioner Henry Mackwilliam, as Mackwilliam recounts in his prefatory epistle, entreated Bedingfield "to afford his paines in the reducing of these few precepts, gathered out of a larger volume written by Claudio Corte, into our English toong."[51]

Corte's "larger volume" comprised three books, the first on the horse, the second on the art of riding, and the third on "the figure of the horseman."[52] When Bedingfield indicates in an epistle dedicating the book to Mackwilliam that "I haue here brieflie collected the rules of horsemanship, according to Claudio Corte in his second booke," he points to *The Art of Riding* as not only a translation and abridgment but also an adaptation: he often speaks of Corte, in fact, in the third person. Bedingfield adds, moreover, that "I haue not Englished the author at large [but have spoken only of those things] that concerne the making of horsses for seruice"; and that he has addressed neither Corte's precepts for "bitting the horsses," nor his "counseling . . . touching the helpe of thee hand," because Blundeville already covered the former and Astley the latter.[53]

While Bedingfield's great contribution was to give an English readership access to Corte's technical precepts on horsemanship, he also introduced Corte's related social ideas on *figura* (appearance, or overall presentation of self). Bedingfield (and implicitly Corte) often advises the rider to avoid (in Bedingfield's English term) "affectation" in seat, while he encourages galloping, for example, as a means of developing a "comelie" seat. Perhaps more to the point, Bedingfield, like Astley, esteems the value of gentleness and denounces the vice of violence more expressly in terms of noblesse oblige than do Grisone (and Blundeville). Citing Grisone's extreme corrections, for example, he admonishes that "to touch [the horse] with fire, or tie chords to his stones, or cats to his taile, as some men doo" is to engage in practices not only cruel and damaging to the horse but also "ouerbase, and vnfit to be vsed by gentlemen." While especially resistant horses might need such "extremeties," he adds, those horses "are vnwoorthie the stable of Princes or Gentlemen."[54]

Bedingfield's focus on making horses for military service, finally, required some finesse. Granting in his epistle that gentlemen typically use horses "more for pleasure than seruice," Bedingfield points out that, nonetheless, "the principall vse of horsses is, to trauell by the waie, & serue in the war." Likewise, he allows that advanced *manège* training enables "Princes & great personages [to] make proofe of the riders excellence [and] to shew the capacitie of the beasts," but he later adds that lower school skills, such as "turning vpon the ground serueth to manie good ends, as well in skirmish as battell, in combate and triumph." He devotes a full chapter to the topic "in what sort you should vse and exercise horsses of seruice for the warre" and advocates, as would generations of subsequent English writers, "Aboue all things you must accustome an horsse of seruice to hunting, where manie other horsses are assembled, and where is great noise and shooting."[55]

With the original publication of Corte's *Il cavallerizzo*, Tomassini observes, "Books on equestrian art [began] to speak to each other." Almost from the inception of the genre, put differently, the discourse on equitation and dressage became not only international but also intertextual: it would proceed through the centuries as an ongoing conversation among and between theorists and practitioners, and masters and students, in books that constantly approved, disapproved, or otherwise referred to one another. Astley's treatise and Bedingfield's translation of Corte provide an almost too literal metaphor for that conversation: the same printer, Henry Denham, not only published both books in the same year, and sold each book separately, but also, and not uncommonly, sold the two books bound as one volume.[56]

Horsemen and Husbandry

Equestrian masters continued to produce trenchant books on horsemanship—in its narrower sense of dressage and equitation—in seventeenth-century Europe, contributing to what Tomassini aptly calls "the slow metamorphosis of the knight-warrior still of medieval style into the modern horseman-courtier."[57] At the same time, new kinds of books began to appear—lengthy and capacious works that expanded the use of "horsemanship" to include farriery and husbandry ("farriery" then meaning veterinary medicine, and "husbandry" horse management). To training and riding horses, in other words, these books added breeding horses and caring for them—the latter comprising everything from shoeing to surgery and, particularly, equine ailments and diseases and preventions and cures for them. Their audience extended beyond courtiers to a broader range of "Gentlemen," particularly gentlemen farmers.

The Perfection of Horsemanship, drawne from Nature, Arte, and Practise (1609), by Nicholas Morgan (15??–16??), is an early exemplar whose key themes and purposes take shape in some twenty-five pages of preliminaries including, in addition to epistles, a commendatory poem, an abstract, and "Admonitions to the Reader."[58] The first three epistles, addressed respectively to the Kings Maiestie, to Prince Henry, and to the Earle of Worcester, Master of His Maiesties Horsse, rehearse the importance of the horse—and thus of horsemanship in its broader sense—to maintaining the military might and political stability of the kingdom.[59] The following epistle, "The Author, to the Gentlemen of Great Brittaine," exhortatory in a way not diplomatically possible in the first three, stresses the importance for "warre and peace" of increasing and preserving horses "in their greatest perfection," and urges its addressees to acquire the requisite "knowledge and practise" by reading books on horsemanship and applying their lessons—advice not simply self-serving, since Morgan invokes Xenophon, Grisone, and Corte.

The Perfection of Horsemanship comprises 186 chapters, the first 56 of them, often theological or philosophical, on the horse and its training and riding. Morgan opens his text proper, in fact, with an ingenious Biblical rationale for an art of horsemanship based on nature: after the Fall, or man's "disobedience," simply put, "now the obedience of all creatures must be attained by Arte," but "no lawfull and humaine Arte can effect any thing against nature."[60] On that premise, he conducts an inquiry into who is fit to learn and teach the art of horsemanship, and what defines the true nature (as opposed to "sundrie diuersitie of natures") of the horse. He eventually concludes "that the naturall inclination of the Ryder, the suffi-ciencie of the teacher, and the nature of the subject [that is, the horse], truly understood with the progresse vse and practise therin, the ful perfec-tion of this Art will be approoued." Masters of that art will triumph "both in Campes and Courts"—that is, in "warre and peace."[61]

Morgan sounds a parallel Biblical note when he opens his subsequent 130 chapters on equine diseases, ailments, curatives, and surgical proce-dures. In accord with prevailing scientific thought, Morgan earlier had explained that the horse, like all creatures, is a compound of four elements, "Fyer, Ayer, Water, and Earth," that issue in four humours, "1. Blood, 2. Fleame [flame], 3. Choler, 4. Melancholy." When the horse was in "the incorrupt state and puritie of creation," the elements and humours were in perfect balance, but "in the degenerated condition of corruption . . . after mans pride had broached the deuils suggestion by the taste of the forbid-den fruite, then appeared the Rebellion of the elements in all creatures." A consequence of the rebellion, equine "sicknesse cannot bee defined other than the disproportion of those foure qualities," so the art of horseman-ship, Morgan reasons, exists to rebalance them.[62]

Morgan's lengthy discussion "of ryding" and training equines aligns so closely with the principles set forth in the previous half century that we need not rehearse them at length here.[63] Morgan's basic reasons for writing a book on perfecting horsemanship, however, do warrant attention. They are twofold: first, "the great decay of good Horses, & the manifolde errors in Horsemanship and the increase of the infinite and intollerable number of Iades [Jades], do so swarm within this kingdome . . . to the dishonor of king and Country"; second, "the true knowledge of Horsemanship has not bene natiue [in England], but onely in forraine Nations." In short, England must develop an indigenous body of knowledge on horsemanship if she is to breed and train "perfect shaped Horses" rather than jades, knowledge whose subsequent growth need not depend on other nations who might "cut all trade and traffick with little England." More than a question of royal honor, then, such knowledge is one of national security and survival.[64]

In sheer number of published works, Gervase Markham (1568?–1637) far outran Morgan and just about every other early equestrian writer. A prolific poet, dramatist, and translator, as well as a writer of nonfictional prose on many subjects, Markham also published a number of books on horsemanship, including *Cavelarice, or the English Horseman* (1607), *Markham's Maister-peece* (1610), and *Cheape and Good Husbandry* (1614), to mention only three. I say "published" rather than "wrote" because Markham (or his publishers) often issued and reissued his work under different titles, in multiple editions, with "repurposed" content.[65] So many were on the market at the same time, in fact, that Markham was "forced to sign an unprecedented agreement with the Stationer's Company" promising to write no more books on diseases and cures of any "cattle" (a common umbrella term for horses, cows, and other farm animals).[66]

Cheape and Good Husbandry carries the subtitle, *for The well-Ordering of all Beasts and Fowles, and for the generall Cure of their Diseases.* While title and subtitle suggest Markham's overall scope and versatility, his volume places emphasis on the horse and its ailments and cures. In an "Epistle Dedicatory to Richard Sackville," Markham invokes his earlier work "as namely of the Horse onely," and promises that he now has "heerin explained a nearer and more easie course for his preservation and health, then hath hitherto been found or practised by any, but my selfe only." The key words are "nearer" and "easie." As Markham writes "to the courteous reader," horsemen and husbandrymen often have difficulty reaching a farrier, and, when successful, often find the farrier lacking "Apothecary Simples." Markham, therefore, will present "those certaine and approved Cures" based on "a few Herbs, or common Weedes [found] in every Field, Pasture, Meadow, or Land-furrow," that will allow the reader to "preserve and keepe his Horse from all suddain extremities"—in short, he gives recipes for home remedies.[67]

Though primarily (or at least nominally) a work of practical husbandry, *Cheape and Good Husbandry* also incorporates Markham's earlier thinking (and likely writing) on horsemanship, "Shewing further, the whole Art of Riding great Horses, with the breaking, and ordering of them," as noted on its title page. Affirming "the Horse of all creatures is the noblest, strongest, and aptest to do a man the best & worthiest services, both in Peace and War," Markham first outlines how to choose "the best Horse . . . according to the use for which you will imploy him." Identifying eight distinct "uses"—"a Horse for the Wars"; "a Horse for a Princes Seat, any supream Magistrate, or for any great Lady of state, or woman of eminence"; hunting; running; "the Coach"; portage; and "the Cart or Plough"—Markham then details the corresponding "ordering of these several *horses.*" With echoes of Blundeville's and Astley's categories for the two basic types of horse, finally, he turns at length to the matter, *Of Riding in generall, and of the particular knowledges belonging to the Art of Riding of a great Horse, or Horse for service or pleasure.*[68]

Markham's commonplace ideas about equitation, unfortunately, lack the inventiveness of his contributions to equine health.[69] The same applies to his ideas on the dressage, or "manage," of the service horse. Advising that "you shall . . . teach him to manage, which is the onely posture for the use of the sword on horse-backe," Markham cautions against "close manage . . . which although it bee artfull, yet it is not so glorious and safe for the Souldiers practice," and advocates for "open manage" and, for example, lateral movements. Like his predecessors, though, Markham equivocates: even the school airs of "close manage" can be "right pleasant and curious to behold, and though not generally used in the wars, yet not utterly uselesse for the same." Like Morgan, in sum, Markham marries the utilitarian value of *manège* riding to the practical value of informed farriery. Together, as he writes in his "Epistle Dedicatory," they "bring a publick good to our Countrey, at which end I ha've onely aymed in this small Book."[70]

In the second half of the seventeenth century, Jacques de Solleysel published his masterwork, *Le Parfait marechal* (1664). William Hope (fl. 1687–1725), deputy-lieutenant and then lieutenant governour of the Castle of Edinburgh, published an English translation, *The Compleat Horseman* (1696), based on the eighth French edition of *Le Parfait marechal* of 1691.[71] *The Compleat Horseman,* according to its title page, would discover "the surest marks of the *Beauty, Goodness, Faults and Imperfections* of Horses," and "The *Signs* and *Causes* of their *Diseases,* the True Method both of their *Preservation* and *Cure,*" together with "The Art of

Shooing, with the several Kinds of *Shooes*," and, not least, "The best Method of *Breeding Colts*; *Backing* 'em, and *Making* their MOUTHS, etc." Since Solleysel "was a veterinarian rather than an equerry," he had a great deal to say about horses, including "the conformation and gaits of the 'ideal' horse," and especially about equine care, but little to contribute to the conversation on riding them.[72]

Hope gets directly to the point in his "Epistle Dedicatory to the King." Since "Your Majesties extraordinary Valour hath brought these Islands once again to the Practice of Arms," Hope's translation "will not only instruct them, to raise a Breed of Warlick and Serviceable Horses . . . but also teach them to train them up, for the benefit of Your Majestie's publick, and their own private Divertisement." His preface then addresses three topics—"the Author . . . his Book [and] this Translation"—and makes three points.[73] First, Solleysel's credentials as a horseman are confirmed by his relationship to "*the Prince of Horsemen*, the Unparalleled and Famous Duke of *Newcastle*."[74] Second, Solleysel's book contains "all that any Gentleman needs to know, either as to *Breeding, Backing, Bitting, Cureing*, or *Shoeing* any kind of Horse, for whatever Service he be designed." And, third, since Solleysel's book contains "little or nothing of the Art of Riding," Hope has added (to cite the title page) "A most Excellent Supplement of RIDING; Collected from the best AUTHORS."[75]

Part 1 of *The Compleat Horseman*, "The Parfait Mareschal, or Compleat Farrier," opens with the ringing assertion, "Amongst all the Creatures, there is none which yeeldeth more profit and pleasure to Man than the Horse," a creature whose many uses include "all the great Interprises of War." Since good horses selected "both for War and the Mannage," however, can become "unserviceable" through their owner's (or groom's) "want of skill in this Art [of farriery]," Solleysel has "taken the pains to make this Book appear in publick, with all the perfection . . . in my power." In the eighty-three chapters of part 2, Solleysel ranges widely across such matters of horsemanship as capably judging a horse's "vigour and agility," foolishly trusting theories based on coloring or humours, and, at more length, properly shoeing, bitting, and breeding (the index alone occupies sixteen pages). His "Discourse on Breeding" mares, finally, belies any notion that the "age of reason" was bereft of romance, "For this Action of Nature," Solleysell observes, "should be performed with Freedom and Love, and not with Reluctancy, and against their Will."[76]

Giants

Literary scholars often measure the importance of a work or its author by the length, breadth, and depth of their historical influence. One cannot overstate the importance, by that measure, of the King James Bible (1539), Shakespeare's First Folio (1623), or Milton's *Paradise Lost* (1667) to early English literature, just as one cannot overstate the importance of Antoine de Pluvinel's *Le Maneige Royal* (1625), William Cavendish's *A General System of Horsemanship* (1658), and François Robichon de la Guérinière's *École de Cavalerie* (1733) to early French dressage—the first two works exemplars of seventeenth-century equestrian thought and the third the epitome of its transformation in the eighteenth century.[77] Though ranging in prose style from the prolix (Cavendish) to the pellucid (de la Guérinière), the three books all feature lavish and illustrative engravings meant not only to beautify, but also to convey detailed information, and they all cover ground from the technical to the philosophical.[78]

The authors of these classics trace their shared lineage not only to the original progenitor Grisone, but also to Grisone's student, the riding master Pignatelli, a key figure in the transition from the Italian school to the French. While Pignatelli published no work in print (some writing did circulate in manuscript), he had two highly influential students who disseminated and refined his ideas: de Pluvinel, who called Pignatelli "the most excellent Horseman of our century as well as of the preceding one," and Salomon de la Broue, who wrote the important and widely circulated treatise, *Le Cavalerice François* (1602). De Pluvinel, in turn, directly influenced Cavendish, and Cavendish, in turn, along with de la Broue, directly influenced de la Guérinière.[79] "With de la Broue and de Pluvinel," Littauer writes, "the fountainhead of equestrian thought moved from Italy to France, where it remained for a long time."[80]

In addition to teaching, training, and writing, moreover, de Pluvinel, Cavendish, and de la Guérinière all held high offices that enhanced their authority beyond the strict boundaries of horsemanship. De Pluvinel, *écuyer* and courtier to Henri III, Henri IV, and Louis XIII, founded the Académie d'équitation at the Tuileries in Paris, where French nobility learned equestrian and other arts essential to their rank. Cavendish, 1st Duke of Newcastle, highly educated and the most intellectually versatile of the three, enjoyed renown as an equestrian and trainer of horses and riders, philosopher and scientist, poet and playwright, music scholar and musician, diplomat and politician, and very wealthy man. And de la Guérinière, the most focused of the three on horsemanship alone and arguably the most influential on the history of dressage, served as equerry to Louis XIV and, building on de Pluvinel's legacy, as Directeur du Manège des Tuileries.

The seminal work of Antoine de Pluvinel (1552–1620) first appeared posthumously in 1623 as *Le Maneige Royal* and in expanded form in 1625 as *l'Instruction du Roy en l'exercice de monter à cheval.*[81] Rather than a treatise, the book takes the classical form of a dialogue between de Pluvinel as master and the young Dauphin (later Louis XIII) as pupil. Part 1, *The King asks Monsieur de Pluvinel what one must do in order to become a perfect Horseman*, as the title intimates, unfolds de Pluvinel's philosophy and theory of equitation and dressage; the much longer part 2, *His majesty begins to ride*, details their practical application. The book rests on the premise, dating to Grisone, that monarchs who learned to judge and control horses demonstrated their ability to do the same with subjects. As the Dauphin says in the opening lines of the dialogue: "I want to learn its [horsemanship's] science and usage, as much as it is necessary to become an excellent Horseman, as well as to be able to judge those in my Kingdom."[82]

A classical humanist of seventeenth-century stamp, as the scholar and translator Hilda Nelson notes, "Pluvinel considers himself not only a teacher of the equestrian art, but also a teacher of virtue and morality." Indeed, teaching—or "schooling"—is the conceptual and structural lynchpin of the two parts of *Le Maneige Royal*: a monarch or nobleman must develop reason, grace, and sound judgment through learning, as de Pluvinel writes, "according to his inclination, his energy, and his disposition," and use them to perfect those skills assigned him by nature and rank; and a monarch or nobleman as horseman (and vice versa) likewise must bring a horse to reason and to the expression of its innate grace through training, according to "his strength, his agility, and lightness, his memory, and his disposition based on either his good or bad will." Perfecting a horseman and perfecting a horse, in short, are parallel processes.[83]

How to achieve the goal of those processes—"perfect harmony between Horseman and horse," or between riding master and horseman (or between monarch and subject)—occupies part 1. "Judgment" is the essential quality for becoming a "perfect Horseman," while "the whole science of schooling horses lies in making a horse obedient to the bridle [left] hand and to the two heels." Using judgment to acquire both grace in the saddle and obedience from the horse demands "patience, industry," and the civility to be "sparing with blows and generous with caresses and flattery." Since "each horse has his own particular air which is natural to him," moreover, the horseman must identify that air and help the horse to express it. As for the particulars set forth in part 2, de Pluvinel advises working a horse "gently, for a short while, and often"; working him "at what he finds most difficult, until he finally obeys"; and working him through a strict regimen of movements, especially the *volte*, "the most important movement a good horse can do, because therein lies more science than in any of the others."[84]

A monarchist who fled England on the execution of Charles I, William Cavendish (1592–1676), 1st Duke of Newcastle, spent his exile training horses and refining the art of *manège*, principally in his riding school in Antwerp.[85] Cavendish published two manuals on horsemanship: *La méthode nouvelle et Invention extraordinaire de dresser les chevaux* (1658), translated into English as *A General System of Horsemanship* (1743); and *A New Method, and Extraordinary Invention, to Dress Horses* (1667)— described by Cavendish as "neither a translation of the first [manual], nor an absolutely necessary addition to it." Though prolix and repetitive, *A General System of Horsemanship* exerted as great an influence in eighteenth-century England as its original version had exerted in seventeenth-century France. De la Guérinière, who invokes Cavendish repeatedly in *École de Cavalerie*, notes that he was "considered . . . to be the greatest expert of his age in the matter of horses." Cavendish, in short, played a pivotal role in the evolution from sixteenth- to eighteenth-century horsemanship.[86]

Cavendish shared many premises with earlier masters, particularly with regard to "vices" and their corrections, but he argued from them to different conclusions. He ascribes the same human qualities of malice, subtlety, and cunning to the horse's disobedience and "his opposition to the rider, in every thing he possible can," for example, just as he warns that a rider responding in anger will not prevail because the horse's "passion" and strength will outdo his. At the same time, though, Cavendish criticizes far more sharply than his predecessors those "ignorant" riders who respond to resistance with harsh methods, particularly when the methods rely on overuse or misuse of spurs and bits. As he puts it, "The more you spur [restive horses], the more obstinately they resist"; or "If [horses] were made tractable by this piece of iron put in their mouths, the bit-makers would be the best horsemen in the world." While the rider should depend on the horse's fear rather than his love, finally, he should exploit that fear not for subjugation but for obedience.[87]

Wary of writers who apply metaphysics and "Natural Philosophy" to horsemanship, Cavendish concerns himself neither with "theological mysteries" of the horse's soul or spirit, nor with divinations of character from coloration or whatever combination of four "elements" is presumed to compose it, "since that is only a piece of empiricism or quackery." Horses do have "natural" capacities and gifts, however, and since "art ought never to be contrary to nature, but to follow and perfect it . . . horses then ought to be worked according to their make, and that form which nature has given them." So, with respect to advanced *manège*, "every horse is to choose his own air, unto which nature hath most fitted him," and the horseman is to enable him to perfect it.[88] As a whole, Cavendish's *A General System* lays out in detail how to do that, with emphasis on "the hand and the heels" as the principal tools "to make a perfect horse," and on the "other things require'd to make him perfectly obedient to the hand and heels"; and on putting the horse "upon his haunches, which is the quintessence of horsemanship."[89]

With François Robichon de la Guérinière (1688–1751), horsemanship moved decisively from the seventeenth to the eighteenth century, not only in period but also in worldview. De la Guériniere's celebrated *École de Cavalerie* (*School of Horsemanship*) first appeared as a complete folio in 1733 (pieces had been published between 1729 and 1731). As Tomassini justly notes, *École de Cavalerie* was "destined to become the veritable Bible of classical riding and [is] still considered the foundation of high-school riding and of modern dressage." It comprises three parts on what its author calls the "three essential aspects" of horsemanship: "knowledge of the horse, the manner of training it, and its care." Part 1 covers topics ranging from conformation to breeds by countries of origin. Part 2, the meat of the work, "contains principles for the dressage of the horse, either for the *manège* or for war, for the hunt or for the carriage." Part 3 catalogs equine illnesses and medications, injuries and surgeries. *École de Cavalerie*, in short, has significant scale and scope, and de la Guérinière, disarmingly modest, freely acknowledges in his preface his intellectual debts for all three of its parts.[90]

Quintessentially eighteenth century, *École de Cavalerie* is encyclopedic rather than didactic and rationalistic rather than moralistic. It reflects and advances core Enlightenment values of disinterested reason manifested in science, system, and method, combined with French values of clarity, conciseness, and elegance in expression. "My design in composing this work," de la Guérinière writes in his preface, "has been to collect and to put into methodical order the principles [behind learned horsemanship]," and "my task has been limited to the development of whatever is true, simple and useful in the art of horsemanship, with a mind to avoiding the tedious excursions and repetitions found in most earlier treatises." As Tracy Boucher has observed, those efforts rest on a conviction in "the perfectibility of Nature's potential through the powers of human reason," joined to "a concept of inherent goodness" in which schooling serves to reveal, or bring out, "the excellent qualities of the horse."[91]

De la Guérinière's thoughts on schooling are too comprehensive and systematic to be captured in a few words, but they can be illustrated with three key points. First, horsemanship must be based on a theory, for "without theory, practical application always remains uncertain"; and the theory, to be effective, must provide "sound principles, and these principles, rather than going against nature, must serve to perfect it with the aid of art." Second, to apply those principles, a "true horseman" must "love horses, have vigour and boldness, and much patience"; must make "knowledge of the nature of a horse"—physical and psychological—"his principal study"; and must acquire "grace" through "a controlled and yet supple posture [maintaining] that exact equilibrium which comes from judicious balance of the body's weight." And third, "*To be between hand and leg* . . . is the quality imparted to a perfectly trained horse"; and "the shoulder-in [is] an indispensable aid in achieving flexibility"; so together they produce "suppleness and obedience, those two principal qualities of a well-trained horse."[92]

De Pluvinel, Cavendish, and de la Guérinière each introduced specific principles and techniques critical to the evolution of horsemanship.[93] Even more important, they collectively advanced a philosophy of horsemanship with the theories that were both its cause and consequence. The view of the horse as an innately malicious and willfully recalcitrant beast, immune to reason, gave way to one of the horse as an intelligent animal mainly prevented from compliance by poor training or riding, one that could be brought to reason with reasoned and reasonable methods. The goal of demanding subjugation gave way to one of commanding willing obedience—not through harsh treatment and tools, especially bits and spurs, but through judicious treatment and simplified tools rather than instruments of torture. And the view of horsemanship as artifice intended to subdue nature, or at least to feign that conquest, gave way to one of horsemanship as an art and, increasingly, a science, developed to perfect nature not with discursive methods but with methodical system. Modern dressage stands on their shoulders.

Denouement

Only three decades later, across the Channel, Charles Thompson (died c. 1790) published a brief book, *Rules for Bad Horsemen, Addressed to the Society for the Encouragement of Arts, Etc.* (1762). Despite its title, Thompson was not writing a lampoon of horsemanship, like his contemporary "Geoffrey Gambado," but a work of useful knowledge for gentlemen who did not claim to be horsemen but who still relied on horses for the road or hunting.[94] Indeed, Thompson added a preface to the third edition of 1765, "lay[ing] no claim to invention: I only offer a collection of [plain] rules gathered from observation"; and an "Advertisement" to the fourth edition of 1775, "tak[ing] the opportunity of this edition to declare, that by *bad* horsemen [the Author] means such whose skill in riding is the meer result of practice, without rules." Underscoring Thompson's view that a bad horseman must gain knowledge of principles (rules, or art), as well as practice, to become a good one, the fifth edition of 1787, evidently the last during his lifetime, carried with its title, *Rules for Bad Horsemen*, the subtitle, *or Those Who Depend upon Practice without Principles.*[95]

In 1765, one J. L. Jackson, Esq. (dates unknown) published another brief book, *The Art of Riding; or, Horsemanship Made Easy: Exemplified by Rules drawn from Nature and Experience.* In a terse notice in the *Monthly Review,* an anonymous reviewer dismissed the book as "Pyratically copied from Thomson's [*sic*] Rules for bad Horsemen; with a few additions, from other publications of prior date."[96] Jackson, in fact, actually led a troop of plagiarists of Thompson, though possibly also of Jackson, that included Charles Hughes, *The Compleat Horseman; or, the Art of Riding Made Easy* (1772) and Laurence O'Reilly, *The Art of Horsemanship, [laying down] Rules whereby Gentlemen may make horses tractable* (1780).[97] That unseemly history of theft aside, Jackson can stand for all four writers in representing

mid- to late eighteenth-century English authors who may have stood on
the shoulders of the French giants (including Cavendish), but who had no
interest in advancing French disciplinary theory or practice. Claiming neither originality nor finesse, they sought to write utilitarian guides for ordinary riders.[98]

The opening sentence of Jackson's preface, indeed, invokes both the
"well known . . . usefulness of a horse" and the little known "right management of him" that results in unskilled riders who render horses "not only
useless, but mischievous." Thus, he will set out useful rules whose practice,
"with care and exactness, will make a perfect and complete horseman,
without any other assistance"—rules that fall into three categories of general horsemanship plus some rules "observed and taught in the manage, or
riding-house; or, How to ride the great horse."[99] Reflecting basic Enlightenment thought, Jackson proceeds to explain that the horse's "natural"
gifts can be harnessed only by "art," so an aspirant first must attain "the
reason [theory or rules] of the art" in order then to combine the "nature,
art, and practice" needed for right management of the horse. Ideally, rider
and horse should seem "as one body in all motions," or "as one piece of
mechanism," or, as Jackson echoes Grisone, Astley, and others, "will do all
things with such harmony and concord of time and measure, that the
spectators will judge the man and horse to be but one body, and to possess
but one mind, one will."[100]

Given Jackson's purpose, he neither urges his reader to seek such
finesse, nor teaches how to achieve it. He does not devalue entirely, however, *manège* riding, and so gives a précis of its principles. He first describes
it as "an art . . . generally considered only as of use to the military gentleman; or to persons of rank," and therefore "quite useless to the generality."
He then shifts ground and allows that in fact it might be useful to general
riders, but decides to beg the question by returning to "our present business [which] is, to give such rules, whereby an unskilful horseman may be
instructed to ride with more safety and ease than, otherwise, he can." A bit
later, Jackson returns to military application, noting that *manège* riding
teaches the trooper how "to make the passage, or side motion, to close or
open the files, and to practise all the military evolutions," then adding:
"But though this discipline may seem peculiarly adapted to the conve-
nience of the trooper, yet may be useful to common riders, when a horse is
given to start or stumble." That hardly counts as a robust endorsement, but

Jackson, in the end, does find some value in school "rules" for even quotidian equestrian activities, although that value, for him, clearly has limits.[101]

Jackson's contemporary, the courtier Richard Berenger (bap. 1719, d. 1782), was appointed Gentleman of the Horse to His Majesty in 1760; he also was considered by Dr. Johnson, for other accomplishments, "the standard of true elegance."[102] In the first of two books, *A New System of Horsemanship, from the French of Monsieur Bourgelat* (1754), Berenger translated Claude Bourgelat's *Nouveau Newcastle, or Traité de Cavalerie* (1747), a book that had distilled, in turn, the Duke of Newcastle's first manual on horsemanship, *La méthode nouvelle et Invention extraordinaire de dresser les chevaux* (1658).[103] Distinguishing between the "*useful* and the *ornamental*" (or *manège*) parts of riding in the translator's preface, Berenger says that among writers on the latter "our illustrious Countryman, *William Cavendish, Duke of Newcastle,* has the highest Claim to our Praise and Acknowledgments." Newcastle's "Want of Method and Exactness," however, prompted the "remedy . . . of the present Undertaking," Bourgelat's "new System of Horsemanship, extracted from the Rules of that great Master."[104] In contrast to Jackson's ornate book on the utilitarian, in other words, this is a useful book on the "ornamental."

A palimpsest of equestrian minds, Bourgelat's *New System* (per Berenger) abridges Newcastle's unshapely treatise into a concise system of succinct principles based on the premise that theory is the foundation of practice, but also is useless without proper execution. These principles include basics of equitation: the "fix'd Point, that just Counterpoise and Equilibre," of a proper seat; and "a *firm* or *steady Hand* . . . [enabling] the just Correspondence between the Hand and the Horse's Mouth." And basics of dressage: "this great Maxim, 'Always observe a just Medium between too indulgent a Lenity and extreme Severity'"; and proper "aids . . . to prevent, and Corrections to punish, whatever Fault [the horse] may commit." The principles also address the proper temperament that must accompany technique. A horseman must display experience, sagacity, forbearance, and patience, and "ought to be Master of the surest Judgment and most consummate Prudence." Only then can he command knowledge and "carry it into Execution," and "teach [the horse] by the Practice of good Lessons to acquire a Facility and Habit of executing whatever you demand of him."[105]

Following Newcastle, Bourgelat also ventured beyond the technical and temperamental to the philosophical. Reflecting an axiom of the zeitgeist, he exhorts the rider "to endeavor to assist Nature as far as you can, by the Help of Art," knowing "that Art serves, and can serve, to no other end than to improve and make Nature perfect." Horses being individuals with differing abilities and inclinations conferred by nature, "a real Horseman will . . . work upon the Understanding of the Creature, [rather] than upon the different Parts of his Body," and will gain "a thorough Knowledge of [the] Horse's Character, [in order to regulate] Lessons and Proceedings conformable to it." For advanced work particularly, an expert horseman first must learn if his horse is "by Nature inclin'd and disposed to the Manage," and then must ken his "particular Disposition, which inclines to some certain Air which suits him best." Whatever the kind or degree of finesse sought, the value par excellence is *harmony*—harmony of the rider's aids, of the horse's motions, and of the two: harmony based on achieving the horse's "Union, without which, no Horse can be said to be perfectly drest."[106]

The second and better known of Berenger's works, *The History and Art of Horsemanship* (1771), comprises two volumes with multiple, variegated parts. Volume 1, in Berenger's précis, includes his recitation of "the history of the equestrian art from its earliest appearance among men, but more *immediately* from its two great sources, *Greece* and *Rome*." Before setting forth in volume 2 "what additions the Art has since received [that is, since Pignatelli], and what the elements are which compose it," Berenger will append to his history, first, "a translation of the treatise of *Xenophon* upon horsemanship," a work notable for its "antiquity" and sole survival, and "still more valuable, as coming from one who as a *General, Historian,* and *Philosopher,* shone with distinguished lustre"; and second, "a dissertation on a *kindred* subject, the *ancient* method of coupling horses in a chariot," by Governor Thomas Pownall. Under the general title, "The Manege," volume 2 comprises a reprint, without attribution, of Berenger's translation of Bourgelat's distillation of Newcastle, followed by "Additions and Remarks" by Berenger and a disquisition on bits and bitting.[107]

If Berenger's *History* now seems conventional, that is because it set many conventions of equine historiography. His opening sentence, for example, invokes "the various and noble qualities with which nature has endowed [the horse]," and his opening discussion affirms that the horse's chief early service lay "in making war" and thus its early ownership likely was restricted to "kings and great men." Following a lengthy enumeration and description of several national types of horse, Berenger unfolds a national history of royalty and horsemanship from William the Conqueror's introduction of horses from Normandy to England, "for the purposes of war, and the exhibitions of public solemnities"; through Edward III's subsequent development of "what is meant at present by a *maneged* horse, or one *dressed* and disciplined for war"—the *magni Equi*, or Great Horses; to the succession of monarchs and their individual contributions to horsemanship, not only including Charles I, praised by no less than Newcastle, but also including the reigning monarch George III.[108]

Formal horsemanship over its first three centuries in Europe ended where it began, as an art of pleasing monarchs by applying the arts of dressage and equitation to winning wars and increasing state power and prestige—an art repaid with ample royal patronage. Master horsemen from Grisone to de la Guérinière had sought to advance horsemanship not only as a discipline of inherent virtue and value, but also as the foundation of mounted warfare, just as master horsemen and horse soldiers like Lewis Edward Nolan or Alexis François L'Hotte would do in the next century. De la Guérinière had the former goal in mind when he wrote, in 1733, of his mission "to revivify the excellence which reigned in the Golden Age of horsemanship." The quest, however, became a more urgent matter of state, certainly in England, with the Seven Years' War concluding in 1763 and the American Revolutionary War and Napoleonic Wars looming. Berenger most likely had the latter—the military—goal in mind when he invoked contemporary British horsemen and horse soldiers, in 1771, and hoped that "under the influence of [his Majesty's] illustrious example, we may expect to see the golden age of horsemanship revive."[109]

2

Manège to Field

Outdoor work is the horse's sole *raison d'être.*

Étienne Beudant
Extérieur et Haute École (1923)

In *The Theory of the Avant-Garde* (1968), the art historian Renato Poggioli argued that acute consciousness of transition—the sense that one is living in a moment of upheaval—generally leads either to nostalgia for a "golden age" or to faith in a utopian future. The long nineteenth century in Europe was such a moment—and was perceived as such by its intellectuals, artists, and equestrian thinkers. While Richard Berenger, in 1771, looked toward a revived "golden age of horsemanship," many of his successors aimed to reinvent classical horsemanship or, more boldly, to invent a new, perhaps even utopian, horsemanship, one that eventually wed *manège* and outdoor dressage and equitation for the training of military horses and riders. Classicists like Berenger held their ground well beyond the late eighteenth century, but romanticists, as in other disciplines, outflanked them in the nineteenth century.

Philosophical currents and political and cultural circumstances provided the backdrop. John Lawrence's *A Philosophical and Practical Treatise on Horses and on the Moral Duties of Man towards the Brute Creation* (1802), for example, illustrates how Enlightenment political philosophy could bear directly on principles of horsemanship. Theories of natural rights, inalienable rights, and human rights associated with John Locke, Jean-Jacques Rousseau, and other progressive philosophers alarmed politically and culturally conservative British thinkers while inspiring their radical adversaries, the brilliant Mary Wollstonecraft prominent among them. Lawrence's chapter "On the Rights of Beasts," his "attempt to vindicate the rights of animals" in general and horses in particular, drew inspiration directly from Wollstonecraft's *A Vindication of the Rights of Men* (1790) and *A Vindication of the Rights of Woman* (1792) at the juncture of Enlightenment and Romantic thought.[1]

That intellectual conjunction, perhaps more to the point, inspired the American and French Revolutions and informed the Napoleonic Wars—circumstances that led to a century of warfare featuring mass cavalry establishments and actions. As Louis A. DiMarco notes, "A phenomenal number of horses, certainly the vast majority of the horse population of central Europe, were directly or indirectly involved in the Napoleonic Wars."[2] That warfare, in turn, informed revolutions in horsemanship, and the broader evolution of the discipline, as much as did conversations among horsemen then ongoing for two and a half centuries. Modern cavalry theory and tactics influenced not only late eighteenth-century equitation, but also the "new method" of dressage subsequently advanced by François Baucher, the rivalry pursued by Baucher and Le Comte d'Aure and the enmity shared by Baucher and Gustav Steinbrecht, as well as the marriage of classical and outdoor riding brokered by François Alexis L'Hotte, François Faverot de Kerbrech, and Étienne Beudant.

Turning the Century

Beginning with Grisone, early writers had promoted the arts of dressage and equitation, in specific and general terms, as being primarily, or sometimes secondarily, applicable to warfare. De Pluvinel, for example, emphasized technical matters of speed, measure, and dexterity in turning because "the horse who turns the fastest and who is solid and sure on his legs, performs best in a duel." To prevail in a duel, moreover, horse and rider must have the timing and skill to speed up or slow down quickly and nimbly in anticipation of their adversary's movement and in reaction to it, or, as de la Guérinière would echo him, "Horsemen of old invented voltes . . . to enable quick turns on the hindquarters, in order more easily to attack the enemy and to escape attack on the hindquarters of their own horses, always being thus head-on with the foe." Though Cavendish also spoke to technical matters, he advanced the broader position that the rigors of *manège* training, in addition to its refinements and contrary to the opinions of ignorant "waggs," not only gave warhorses endurance for long marches, but also made them "much fitter for galloping, trotting, wheeling, or any thing else which is necessary."[3]

Mobile and versatile light cavalry soon would supersede slower and more constrained heavy cavalry, and charges at speed and related tactics would moot duels fought essentially in place. In that context, de la Guérinière devoted a chapter of his *École de Cavalerie*, "On War Horses," to the qualities and training of cavalry mounts. Suitable horses, he writes, must be neither vicious, "for in the heat of battle the rider would not be able to control [them]," nor intractable in overcoming natural fears, since they will encounter many fearsome sounds and sights in battle. Horses educated in only one of the "two sorts of manège: military school, and that of the arena, or high school," he came to believe, moreover, would gain from both: "The art of war and the art of horsemanship owe to each other many benefits. . . . The use of [school] principles has contributed to the precision of the various manoeuvres used in armies"; and even "if the airs above the ground offer no advantage for war, they offer at least that of imparting to the horse the agility it requires to clear hedges and ditches."[4]

As for horses, so for riders. From the beginning of formal horsemanship, the discourse on training horses proceeded hand in glove with that on training riders. The latter emphasized the equitation skills needed not only to manage and navigate a spirited warhorse both over long marches and in the chaos of battle, but also to fight effectively—with lance, saber, or firearm—while mounted. Without skill in *manège*, Cavendish wrote, "one cannot be a good horseman, nor ride a horse boldly, either for pleasure, or in war."[5] Accordingly, the late eighteenth century witnessed the beginning of many decades of writing on schooled equitation and its application to the military in general and to light cavalry in particular. English thoroughbred horses and cross-country riding as practiced by generations of foxhunters figured in this conversation, so English writers joined in; as a debate on *manège* principles and practices, however, it was largely a Continental affair.

How German is it

The French, of course, held no monopoly on equitation in this period, as Werner Poscharnigg makes clear in *Austrian Art of Riding* (2015), his study of "the essence of Austrian equitation" over five centuries. Poscharnigg traces that "essence" to Sigmund von Josipovich's doctrine of "the horse's swinging, elastic gait and the rider's pliant, elegant seat in harmony with it," and he identifies its key elements as "the thinking rider, the largely humane treatment of the horse, the emphasis on effortless riding, even for the cavalryman"—elements that translate, in technique, to "a light hand," or "the cessation of the aids once they work, the *descente de main*." While Poscharnigg gives Baron von Eisenberg due respect in this history, he names Johann von Reganthal the seminal figure, attributing to Reganthal not only the concept of the "thinking rider," guided by sound reasoning, but also the beginning of "a new era of effective and humane handling of the horse," as Reganthal elaborated in the "compendium" of his experience, dictated just prior to his death in 1730, discovered in manuscript in 1996, and subsequently published as "The Primary Directives."[6]

Specific features of Austrian equitation aside, Poscharnigg locates its evolving "essence" in its being inherently and integrally martial. After some two centuries, he argues, European equitation, particularly in pre-Revolutionary France, had diverged from military use to a point where "a hyper-refined equitation . . . alienated itself more and more from the battleground" and became, in his view, "art for art's sake." The post-Revolutionary French army, as a result, rejected *manège* equitation as "anti-republican." The high school elite in Vienna, by contrast, "passed on its scientific principles to the military," a rapprochement that "preserved the High School from artifice," atrophy, and extinction. It also led to a long-term and mutually beneficial reciprocity of school, campaign, and terrain riding whose apotheosis was Major General Leopold Freiherr von Edelsheim-Gyulai's *Exercise Manual for the Imperial and Royal Cavalry* (1874), "a monument of superior equestrian culture," a work whose value, Poscharnigg argues, lies as much in its philosophical assumptions and implications, both humanist and romanticist, as in its technical specifications.[7]

One of the figures in the evolution traced by Poscharnigg, Friedrich Wilhelm Baron Reis von Eisenberg (1685–1764) was the author of *The Art of Riding a Horse, or Description of Modern Manège, in its Perfection* (1727). Elegant in conception and execution, *Modern Manège* offers, as its English translator notes, forty-eight lessons in "how to train a horse properly to perfection in *Manège*." Each lesson comprises a page of text and a facing illustration, expertly drafted and engraved, the former written by Eisenberg and the latter drawn by him (and engraved by Bernard Picart). Eisenberg intended to reach "Connoisseurs" and "Apprentices" alike by addressing with only "a minimum of words the essentials of the necessary lessons," and by focusing instead on "giving [the reader's] eyes a chance to see the correctness of these Manège horses in their actual situations." Eisenberg, in short, wanted to *show*, not tell.[8]

Modern Manège opens with seven descriptions and illustrations of equine types (Spanish, Barb, Neapolitan, and so on) "according to their size, dispositions, and different aptitudes" and the training methods therefore most appropriate to each. Eisenberg follows these with the before-mentioned forty-eight highly systematic lessons, cautioning that "the Art of Riding a Horse demands that one lesson follows another precisely." The lessons advance through the basic gaits, passage (half-pass) and other suppling exercises, airs on the ground, and, finally, airs above the ground. Yet however sophisticated these movements may become, Eisenberg writes, "the trot is the foundation of the Manège work. . . . Everything emanates from it." Offering each lesson both with the bridle alone and with "the incomparable effects of the cavesson," Eisenberg concludes with seven bit types and their applications—as did, with varying numbers of bits, many of his predecessors.[9]

Four motifs recur in the lessons. One, echoing Cavendish, horses possess natural aptitudes that the trainer must identify, accommodate, and enhance: nature disposes "jumpers" for airs above the ground, for example, while trainers only "give them time." Two, horses do not respond to harsh or impatient training, but demand and deserve "reason and patience," or "softness and discretion, and light aids," or "kindness and caresses." Three, airs are beautiful—the capriole, for example, is "the most beautiful, but also the most difficult of all the airs above the ground. It is also the rarest because there are few horses which are capable of doing it." Four, as noted in the translator's introduction, "the picture of the horse when ridden should not be marred by an ungraceful rider": Eisenberg himself finds nothing "uglier than to see a rider who lets himself move back and forth" or "who [holds] on with the calves of his legs." Eisenberg also had introduced a key point—and a bald appeal for patronage—in his dedicatory letter to George II: "The Art of creating riders and horses according to the rules of the Manège is . . . an indispensable necessity for war."[10]

Early eighteenth-century contemporaries of de la Guérinière, Eisenberg and Reganthal had many important legatees in the next century, not the least of them Maximilian von Weyrother (1783–1833). Chief Rider of both the Campaign Riding School and the Spanish Riding School in Vienna, Weyrother introduced the work of de la Guérinière to the latter, and was "dedicated to the installing, practising and promoting of de la Guérinière's principles and equestrian legacy," in the words of Daniel Pevsner. Weyrother published only one minor work during his lifetime, a short treatise on bitting with the very long title, *Anleitung wie man nach bestimmten Verhältnissen die passendste Stangen-Zäumung finden Kann: nebst einer einfachen Ansicht der Grundsätze der Zäumung* (1814); his pupils, however, assembled and published his papers as a slim posthumous volume, *Bruchstücke aus den hinterlassenen Schriften des k. k. österr. Ober-bereiters Max Ritter von Weyrother* (1836), translated much later into English as *Fragments from the Writings of Max Ritter von Weyrother.*[11]

Despite Weyrother's regard for de la Guérinière, the *Fragment*s introduce a new, Teutonic premise illustrative of the different approaches taken by the classical French and Austrian traditions: although a horseman "must be aware of scientific principles," Weyrother proposes, dressage and equitation "cannot create [their] own principles . . . but must borrow them from elsewhere." Weyrother bases his signal insight on that premise: "The first principle of the art of equitation is taken from statics, the general principle of equilibrium, that is the more the line of force of the centre of gravity falls into the base of the body, the more stable the body will be"—a principle that he then clarifies: "The underlying rule of equilibrium is as follows: the lower the centre of gravity, or the shorter the line of force in relation to the diameter of the base, the more stable will be the body." Weyrother's method rests wholly on this principle from mechanics.[12]

Since mastery of riding is prerequisite to mastery of training horses, and since proper riding presupposes a proper seat, Weyrother starts there in applying his principle of equilibrium: "The vertical position of the pelvis therefore determines the whole posture of the rider, and is thus the basic principle for the seat on the horse. . . . All instruction must be based on this established basic principle and constantly refer back to it." Weyrother's pupils arranged the remaining fragments of his writing, therefore, to advance from teaching a novice rider to mount a horse to training a green horse to serve. With respect to the latter, Weyrother's principle, "All resistance that the horse undertakes stems from fear of man, or from ignorance of what man wants of the horse, or alternatively from inability to perform what is required"; and his inference that instilling "trust" is therefore the key to training marks his distance from harsh early writers who attributed resistance to equine perversity or maliciousness. His pupils, though, whitewashed an apostasy that employed not only whip and pain, but also hunger and thirst, to tame "a horse that can no longer be corrected by kindness."[13]

Weyrother's pupils appended to these fragments, finally, a proposal that he had made for a military institute. Weyrother's methods in general, they pointed out, would lead not only to an increase in military equestrian skill, "but also to a substantial reduction in remount costs through appropriate training of the horse," and his institute, moreover, would serve "to achieve uniform practice of equitation in the imperial and royal cavalry." Specifically, it would use school horses to train riders who then could train school horses—training not only in movements for the *manège,* but also for performance "in the open field . . . jumping over ditches and fences." For Weyrother, as for others in the Austrian and German traditions, *manège* and cross-country training were not opposed. In the case of exceptional horses, as Poscharnigg paraphrases Weyrother, the former "crowned" the latter. In this, Weyrother marks his proximity to the most prominent French horsemen of mid-century.[14]

Baucher and Company

Historians agree on the trajectory of Continental horsemanship from the late eighteenth century to the early and later nineteenth century: France dominated classical equitation in the eighteenth century; the French Revolution of 1789–1793 denigrated the discipline as overly refined, aristocratic, and anti-republican; and "Napoleon and his wars accelerated [its] decline." Once revived, classical equitation in France, as in Austria, also had to evolve to serve not only mounted military in general, but also, more specifically, vast cavalries fielded by modern European armies. Chenevix-Trench legitimately could contend, in *A History of Horsemanship,* "From 1815 to the end of the century France, and particularly the great Cavalry School of Saumur, was the home of educated horsemanship," but French horsemanship in the new century also brought into relief previously latent links between classical French equitation, German cavalry equitation, and English cross-country equitation. This triangulation would define European military horsemanship going forward.[15]

At the circus

The brilliant rider and theoretician François Baucher (1796–1873), as controversial as he was influential, stood at the center of nineteenth-century equitation. Following his *Dictionnaire raisonné d'équitation* (1833) and *Dialogues sur l'équitation* (1835), Baucher's chef d'oeuvre, *Méthode d'équitation basée sur de nouveaux principes* (1842)—often called simply the *Nouvelle Méthode*—was published in eleven editions between 1842 and 1863, and in a revised twelfth edition, published in 1864, that introduced what is known as Baucher's *deuxième manière* (Second Manner or Second Method).[16] The book and Baucher's riding exhibitions prompted contemporary attacks by Louis Seeger and P.-A. Aubert, among others, accusing Baucher of being "less an innovator than an 'adjuster' of various defective procedures put into practice for the corruption of equestrian art."[17] Baucher also attracted twentieth-century encomia, however, by General Decarpentry and Nuno Oliveira, among others, declaring Baucher the most innovative theorist since de la Guérinière (Oliveira) or, more extravagantly, since Xenophon (Decarpentry).[18]

Baucher's "new method," in fact, not only provoked contemporary attacks from his adversaries and defenses from his allies and disciples, but also has prompted surely more commentary and explication in the twentieth- and twenty-first centuries, by partisans and neutral observers alike, than any other nineteenth-century equestrian figure.[19] These commentators, in a manner of speaking, responded to the call for guidance in Felton's uncharitable assessment of the *Nouvelle Méthode* as "being so obscure as to be almost inevitably misunderstood by anyone attempting to follow Baucher's text without any further guide to his meaning."[20] Despite Baucher's alleged obscurity, the considerable nuances of his system, and the varying points of view and emphases of his commentators, the latter in fact share a high degree of consensus on Baucher's basic goals, principles, procedures, achievements, and limitations, as well as on critiques leveled against him:

Though a brilliant rider, Baucher attended primarily, though not exclusively, to horses: his system is one of dressage rather than equitation. "Baucher's method is almost exclusively devoted to the schooling of the horse, less to the training of the rider," Nelson notes. "Only much later did he include a chapter dealing with the schooling of the rider."[21]

Baucher's system assumes two forces: instinctive force, when a horse uses his muscular action as he chooses, and transmitted force, when the action is produced by the rider. "Once the horse has been mounted he should move only as a result of *imparted* [transmitted] force," as Littauer explains. The first aim of the rider, then, is to "destroy the *instinctive* forces" and the second is "to replace the *instinctive* forces by *imparted* ones."[22]

The *effet d'ensemble* (or "coordinated effect") provides the primary tool for annulling the instinctive forces. Dom Diogo de Bragança writes, "It consists of the simultaneous use of the impulsive aids [leg] and the holding aids [hand], in a fashion that the opposition of forces obtained by those aids leads to the complete annulment of these forces." This tool, as Baucher realized, and as Hilda Nelson explains, proved invaluable when training "the problem horse . . . with a poor conformation."[23]

Having annulled the instinctive forces by the *effet d'ensemble*, the rider takes possession of the horse's strength, action, and will, and the horse submits: "The Old School horse in his *submission* remained haunted by the fixed idea of forward movement," Decarpentry writes. "A Baucher horse had no other concern than to strictly conform to received orders, whatever their demand or direction, without ever anticipating them, nor, above all, exceeding them."[24]

That possession and submission enable the rider's use of partial actions to achieve total lightness (that, reciprocally, benefits those actions). "The *partial actions*, a fundamental point in Baucherist dressage," de Bragança notes, "consists of relaxing (decontracting) a chosen part of the body without letting other parts of the body participate in the executed movement"— or, put differently, Baucher supples the jaw, poll, spinal column, and so on *separately*.[25]

Suppling and flexion, in turn, enable achievement of *ramener* (or vertical positioning of the head), and *ramener*, in turn, enables achievement of *rassembler* (essentially, collection), though the two, more properly, are causes and effects of one another. Baucher departs from the classical tradition, however, in his belief that collection precedes impulsion rather than follows from it.[26]

As a consequence of that belief, Baucher commits the heresy of treating the walk, rather than the trot, as the basic gait, "And that is understandable," de Bragança allows, "because a horse does not submit to the *effet d'ensemble* perfectly except in place or in a slow walk." Committing an even worse heresy in the eyes of his adversaries, moreover, Baucher prepared for work at the walk "by a series of exercises asked for *in place.*"[27]

In the end, Baucher's "absolute subjugation of the horse" carried a cost: "it could provoke *acculement* (backing up, sucking back, and coming behind the bit) or at least 'extinguish' an animal that had fire and natural impulsion," turning a Baucher horse, as Aubert put it, into an "ambulatory cadaver." Baucher's critics damned him, in general, for ignoring "the morale, will and willingness of the horse . . . and totally subjugating the horse to the will of the horseman."[28]

Baucher's "Second Manner" modified rather than nullified those principles, though commentators debate why and how. For General Decarpentry, for example, "this 'manner' is only the last of the perfections that the Master had continuously brought to his work over the course of his long career. . . . All the essentials of the method, its principles, its spirit, remain intact [including] the '*mise en main*' [lightening of the forehand] and the '*effet d'ensemble*' . . . but the means to obtain each are modified in their nature, their combination, and the measure of their application."[29] The crucial difference lies in the shift from *effets d'ensemble* to the Second Manner principle of "*main sans jambs, jambs sans main*" (hand without legs, legs without hand): rather than hand and leg working simultaneously to annul instinctive force and establish transmitted force, the hand and leg work in coordination, the leg giving impulsion and the hand direction and lightness.[30]

Needless to say, Baucher himself addressed all the points that his commentators have glossed (save those concerning the Second Manner) as he moved systematically in earlier editions of the *Nouvelle Méthode* through the rider's seat, the horse's forces, the means for suppling the horse, and the rider's employing and concentrating the horse's forces, concluding with a kind of executive summary, "Succinct exposition of method by Questions and Answers":

"The horse," Baucher opens his discussion of forces, has "a weight and a force peculiar to himself," and, "from the moment he is mounted, should only act by *transmitted forces.* The invariable application of this principle constitutes the true talent of the horseman."[31]

"A horse puts himself in motion only in consequence of a given position; if his forces are such as to oppose themselves to this position, they must first be annulled, in order to replace them by the only ones which can lead to it."[32] Proper motion presupposes proper position, and proper position presupposes the annulment of conflicting instinctive forces and their replacement with transmitted forces. The *effet d'ensemble* is the tool.

"The education of the horse consists in the complete subjection of his powers [and] we can only make use of his powers at will, by annulling all resistances." Once the rider has eliminated resistance and taken control of the horse's powers, "the animal in all his movements should do nothing of his own accord."[33]

Since resistances result from contractions in separate parts of the horse, such as jaw and poll, the rider should "combat them separately"—that is, supple them—with partial actions and thereby attain the true goal: "Lightness, always lightness! this is the basis, the touchstone of all beautiful execution." Thus, on the one hand, "Violent effects of force should be avoided, for they only bewilder the horse and destroy his lightness," while, on the other hand, equilibrium of the forces of a horse in motion must be maintained: "If equilibrium is only obtained by lightness, in return, there is no lightness without equilibrium."[34]

If lightness is the goal, and if collection precedes impulsion, then "the difficulty of horsemanship does not consist in the direction to give the horse, but in the position to make him assume." *Ramener* and *rassembler* are the keys to that position. Horsemen have mystified the latter and ruined horses as a result, however, leading Baucher to proclaim, "The gathering a horse [collection] has never been understood or defined before me, for it cannot be perfectly executed without the regular application of the principles which I have developed for the first time"—a claim, needless to say, that only fueled the enmity of his critics.[35]

Finally, and succinctly, the walk "is the mother of all the other paces; by it we will obtain the cadence, the regularity, and the extension of the others"; and the "only means of obtaining precision and regularity of movement in the walk and the trot, is to keep the horse perfectly light."[36]

Baucher begun the opening chapter of all but the earliest editions of the *Nouvelle Méthode*, "Obtaining a Good Seat," by addressing what might seem the "astonishing" fact that he had not begun "by speaking of the rider's seat," something so important that it "has always been the basis of classical works on [horsemanship]." He explains, far from modestly, that he "wished to make a thorough reform," but since his "system for giving a good seat to the rider" also was "an innovation," he did not want to overwhelm amateurs or affront adversaries with concurrent innovations in dressage *and* equitation. He now advises instructors to emphasize the fundamentals of rider position "to explain all to the rider, and enable him to obtain a good seat," and, at the appropriate later point, the fundamentals of horse training, so the pupil will understand "what an intimate connection exists between the education of the man and that of the horse."[37]

Developing two key points about equitation, Baucher, one, assigns all blame for faulty development of a horse to the rider, and two, attributes final success for correct development of a horse to equestrian "tact." With respect to the former, he argues that resistances have physical causes (primarily faults in conformation) and therefore demand physical solutions prevented only "by the awkwardness, ignorance, and brutality of the rider." Or, as he places first in his summary questions and answers: "Is it to the rider or to the horse that we ought to impute the fault of bad execution? To the rider, and always to the rider." With respect to the latter, he makes clear that the proper rider, by contrast, has correct position, a light hand, and, above all, "a fineness of touch, a delicacy of equestrian feeling . . . without which, we seek in vain to pass certain limits"; that "fineness of touch, that delicacy of process, indispensable to the right application of my principles"; or, in a word, "tact."[38]

Baucherists and daurists

Baucher and his methods stood at the center of an intranational controversy that played out polemically in the microcosm of Parisian equestrian society. De Bragança reminds us that when the distinguished horseman Le Comte Antoine Cartier d'Aure (1799–1863) became *écuyer en chef* of the École de Cavalerie at Saumur in 1847, "he put academic dressage to the side, and made a more practical method prevail, [one] in his view more suitable for a military school." With that method, Felton writes, d'Aure sought "to develop a cross-country and jumping seat based on classical principles." The virtues and vices of Baucher's and d'Aure's systems, respectively, of high school *manège* riding and *l'équitation extériur* (outdoor riding) were the subject of an extended public debate with influential vocal partisans—*baucheristes* and *dauristes*, as they were known—in formation on both sides. The stakes were not trivial.[39]

Given the critical role played by cavalry in mid-nineteenth-century Europe, and given the hegemony exercised by the École de Cavalerie over the training of horses and men for the French cavalry, an ambitious French *écuyer* with a system that he considered revolutionary, like Baucher, would want to establish that system in the military to ensure its prestige and legacy.[40] Baucher, however, faced at least three obstacles. First, as Nelson notes, "the aristocracy and the *haute* bourgeoisie" controlled horsemanship; they also controlled the cavalry that controlled military horsemanship. From the petite bourgeoisie, Baucher was neither an aristocrat nor a high bourgeois, nor was he a cavalryman. Second, Baucher developed and demonstrated his theory and practice in the circus—primarily Laurent Franconi's Cirque Olympique. Although the circus was a respectable venue for dressage and equitation, it was not a military venue. And third, Baucher dedicated himself to school riding indoors, so "his methods were of little value as applied to the outdoor horse."[41]

Baucher's aspirations came to a head in a famous public challenge in 1842. Pitting his "new method" against that of his rival Le Comte d'Aure, Baucher led a previously unmanageable thoroughbred, Géricault, through a precise and correct routine after the briefest training.[42] Everyone who mattered in Paris attended, *dauristes* and *baucheristes* alike, including Baucher's most important advocates, the Duke d'Orléons and General Oudinot. Largely as a result of this tour de force, the military conducted tests of Baucher's method that led, in turn, to a course at the École de Cavalerie in 1843 for training officers in that method "under the direction of Lt. General de Sparre" and the oversight of Baucher. On the basis of its progress, and the testimony of participating officers, Baucher had good reason to believe that the military would adopt his system. To his disbelief, the Comité de la Cavalerie, the body establishing policy for the École, canceled the course in 1845 and rejected Baucher's method, citing a military rationale and leaving its political reasons unspoken.[43] Though d'Aure lost the first skirmish, in a word, Baucher lost the battle.

"The thinking of both [Baucher and d'Aure]," Littauer explains, "had its roots in the 18th century," with Baucher transforming classical dressage to fit the needs of training military horses, and d'Aure adapting classical equitation to meet the demands of outdoor military riding. Classically trained, d'Aure had come to believe that *manège* schooling could prepare horses with the suppleness, lightness, and obedience needed for outdoor riding, but also that, alone, it could train neither horses nor riders for outdoor riding adequate to military purposes.[44] Influenced by English civilian riding and German military riding, d'Aure diverged from Baucher on many issues, including impulsion or the disposition of the horse's "force." D'Aure, Littauer notes, "was particularly critical of Baucher's theory of the destruction of the natural forces of the horse," and instructed riders, by contrast, as Nelson writes, "to leave the horse his natural energy . . . to place the horse in the proper position and allow him to execute the movement on his own."[45] Hence d'Aure's slogan, "Forward! Forward! Always forward! And once again forward!"

The mythology of artistic and intellectual rivalries—particularly those in modern Paris—typically involves an apocryphal meeting, some begrudged signs of mutual respect, and an outcome of even greater benefit than expected—in this case, respectively, Baucher and d'Aure reputedly met once to negotiate the sale of a mare; Baucher acknowledged d'Aure as a natural "centaur," and d'Aure never attacked Baucher *personally*; and, in the end, their controversy "contributed to the prestige of the two equerries and to the return of France's prestige in the field of equitation in general." While "it is quite obvious that Baucher, with his inclination toward the artificial, and the Count d'Aure, with his pursuit of simplicity and naturalness, could never agree," as Littauer writes, their deeper difference may lie not in horsemanship per se: Baucher embodies the intellectual dedicated to perfecting a theory and demonstrating its correctness, while d'Aure resembles the artist or craftsman committed to perfecting a practice whatever its theory.[46]

Anglomaniacs and romanticists

Two broad and related cultural phenomena framed the Baucher versus d'Aure rivalry: Anglomania and Romanticism. Anglomania in the French equestrian community had two aspects: one, a vogue for breeding and training English thoroughbred horses; and two, a vogue for riding them in the English cross-country style at the gallop, as well as in the *trot d'Anglais* (or rising trot).[47] Romanticism as pertains to that community likewise has two aspects, but of a different order: one, effects of the Romantic zeitgeist in general, and Romanticism in the arts in particular, on actual equestrian pursuits; and two, our current academic debates about the kind and degree of those presumed effects.

With respect to the question of Anglomania, Mike Huggins has argued that "the Civil War's damage to British breeding stock [in the seventeenth century] created a perceived need to 'reinvigorate' British military and riding horses," and that the eighteenth-century foundational Arab stallions and mares met a functional need while providing a symbolic value. He connects (as have others) the aristocracy's parallel obsessions with both equine and human bloodlines: "Belief in the superior breeding of their horses became linked to belief in their own superior breeding." Thus, Huggins argues, the thoroughbred became "a figuratively powerful cultural icon within the British imagination," and the self-styled "British creation" of the thoroughbred "helped to sustain the wider belief in British superiority through the war with France and the loss of America." Ironically, those same English thoroughbreds were in vogue and high demand in France following the Napoleonic Wars, also for reasons of prestige as well as for their quality and traits.[48]

New systems of horsemanship often follow new breeds of horses. "The basic reason for the modifications brought to the methods in use [in French dressage]," for example, as de Bragança points out, "was the new vogue for thoroughbred horses, who served often as studs for crossbreeding." French equitation, historians concur, followed suit, amalgamating English outdoor riding—derived primarily from foxhunting and steeplechasing, and much admired for its boldness if not, indeed, recklessness—with *manège* riding for both military and civilian purposes. Tracing a similar phenomenon in Austrian equitation, Poscharnigg makes the connection to Romanticism: "[The] revolt against the frame of rational equestrian rules corresponds to the 'Sturm und Drang' period, a youth-movement since 1750 that wanted to have rational rule exchanged for the unleashing of strong sentiments, freedom, heart, desire, and overwhelming feelings."[49]

With respect to Romanticism, as noted earlier, one can probe how the general liberal and revolutionary zeitgeist in Europe from roughly mid-eighteenth century to mid-nineteenth century may have influenced horsemanship, or how the specific Romantic movement in philosophy, literature, art, and music that emerged from that zeitgeist, and that is associated with nature, freedom, and rebellion against norms and constraints, may have influenced it. Applying the question to Baucher, de Bragança, in a brief discussion of "Baucherism and Romanticism," advances the reasonable proposition that the zeitgeist likely enabled and encouraged not only Baucher's innovations but also his conviction that his innovations were revolutionary. Nelson, in a lengthy discussion of the same topic, takes a more tendentious, though not indefensible, position: "The battle between Classicism and Romanticism that affected literature, art and other fields of endeavor was also evident in the art of riding."[50]

The problem lies in the application of terms. First, attributes assigned to Classicism and Romanticism in the arts do not map cleanly onto attributes of horsemanship. Second, Classicism in horsemanship may mean the tradition culminating in de la Guérinière, but it does not follow that rejecting it makes Baucher (or anyone else) ipso facto a Romantic.[51] And third, Classicism and Romanticism in any context remain malleable terms that can be opposed on multiple and unaligned axes. So, for example, d'Aure represented the establishment and Baucher the rebel outsider, so Baucher gains an attribute of Romanticism; d'Aure, however, advocated outdoor, "natural," very bold, and in some sense "simple" horsemanship, while Baucher advocated indoor, artificial, and highly controlled and sophisticated horsemanship, so d'Aure gains the Romantic advantage.[52] In other words, Nelson may defend the position that "the Classical and the modernist [meaning Romanticist] were epitomized [respectively] in the methods of equitation of Baucher and D'Aure," but, in the end, she cannot hold it intact.[53]

The Franco-Prussian (Dressage) War

France, of course, is not Germany—in horsemanship as in many other things—and particularly was not in the middle third of the nineteenth century. While the heterodoxies of Baucher and d'Aure contended for supremacy in Paris, classical dressage orthodoxy based on de la Guérinière held firm in Berlin. Baucher and his methods, as a result, stood at the center of an international (as well as intranational) controversy—as the bêtes noires of German dressage theory and practice. Already incipient in the eighteenth century, as Tomassini has noted, the clash between the French and German schools, though certainly less consequential than the Franco-

Prussian War of 1870, nonetheless proved more than a local, nationalistic squabble.[54] It would inform the future of both dressage and cavalry training in the twentieth century.

As often happens in equestrian history, sixteenth- through eighteenth-century Italian and French equitation being the obvious example, the German side of the nineteenth-century French–German dispute follows a genealogy of teachers and pupils. Maximilian von Weyrother's pupil Louis Seeger (1798–1865), the distinguished German dressage master, theorist, and devout follower of Classicism and de la Guérinière, was the author not only of *System der Reitkunst* (*System of Horsemanship*) (1844), but also of *Herr Baucher und seine Künste—Ein ernstes Wort an Deutschlands Reiter* (*Mr. Baucher and His Arts: A Serious Word to Germany's Rider*) (1852). Seeger's star pupil and protégé Gustav Steinbrecht (1808–1885), in turn, was the author of the brilliant, posthumous, and anti-Baucherist treatise, *Das Gymnasium des Pferdes* (*The Gymnasium of the Horse*) (1886), a landmark in German dressage theory.

Three motives initially drove the polemic that Seeger wrote to save German riders from being gulled, like the French public in his view, by the charlatan Baucher—namely, Baucher's arrogance, celebrity, and illusionism. Baucher, that is, "[has presented] his own method as . . . the only salvation of equestrian art that now lies in ruins," and "has posed himself as founder and prophet of an entirely new method that must be the one and only"—an unseemly arrogance that did not suffer public opprobrium, but, to the contrary, enabled demonstrations and works that made Baucher a public sensation and "the talk of the whole world," that is, a celebrity. Baucher may be only a showman with a "talent for presenting his horses while *conjuring tricks* likely to impress the crowd" (my italics), but his visibility and popularity "[demand] that we make a serious examination of his work."[55]

Because Seeger's goal was to debunk Baucher's method rather than to advocate his own, his examination devotes much attention to alleged faults, although, as de Bragança correctly cautions, "Seeger's criticism dates from 1853 [and] Baucherism, at that time, had not attained the maturity that would lead it to the Second Manner." In theory, for example, as Seeger wrote, Baucher's method is "a complete and closed system" that effectively precludes elimination of any of its parts. In practice, it results in "bascule movement," painful to watch; in throwing "the rider's weight onto the forehand"; in "evasion behind the reins"; in the horse's demoralization; and, as its goal, in "removing the spirit of resistance in the horse." The indictment, moreover, extends beyond the method to its creator: "M. Baucher uses the whip as punishment, with an extraordinary severity," and the whip and spur "to an extreme degree that borders on cruelty."[56]

Seeger's polemic, however, does not comprise simply a bill of particulars: it *is* a "serious examination" of Baucher's foundational principles in opposition to Seeger's own. For Seeger, Baucher's method fundamentally represents "a transgression of the rules of nature and art": weakening the horse's "natural tendency to forward movement," it is "a preparation . . . that is against nature." While Seeger and the German school borrow "our rules . . . from the very mechanism of the horse's movement . . . M. Baucher does not comply with this natural law of movement." Baucher, in essence, has raised a defective idea "to the level of a principle [and] erected [it] into a system, the *impairment of the natural impulsion of the horse*," whereas "our principle is: *strengthen the horse in movement for the movement*." Not surprisingly, Seeger concludes, Baucher's horses are unusable "for the hunt or cavalry service," and "his method [was] abandoned after a short period of practice in his country's cavalry."[57]

Though far more aggressive than even Seeger's indictment, Steinbrecht's assault on Baucher in *The Gymnasium of the Horse* remains only one element in a rigorous work of great scale, scope, and significance—and, incidentally, of a complicated "authorship" that includes Steinbrecht, his pupil and editor Paul Plinzner, and Plinzner's pupil Hans von Heydebreck.[58] William Steinkraus, the late giant of international show jumping, asks in his foreword to a recent edition: "Who was the author of this book the Germans consider worthy of ranking with Xenophon, Pluvinel, Newcastle, and Guérinière?" The answer to that question—or at least the answer of Daniel Pevsner, late of the Spanish Riding School—is this: he was the "father of modern German equitation."[59]

Steinbrecht's treatise, its editor Plinzner wrote, was neither "intended for the layman" nor written as "a guideline for training horses. . . . Instead it attempts to immovably set the goal of the equestrian art, to show what means the horse's nature gives us to attain these goals, and to explain how a system of gymnastic exercises that we call the training of horses can be assembled through appropriate use of these means."[60] That concise précis nicely captures the essence of a lengthy and lucid exposition that unfolds in five topical sections: the rider's seat and aids, the purpose of dressage, systematic training of the horse, school movements, and an epilogue. Though detailed and densely interwoven, Steinbrecht's basic principles for the gymnastic training of horses emerge with clarity from that exposition.

Though Steinbrecht immediately establishes his object as "the *working of the horse*," for example, he begins *The Gymnasium of the Horse* with the training of the *rider* who must do that working. Everything in riding, he repeats often, depends on "a perfect seat and fine feeling." In order to develop soft hands, in Steinbrecht's view, the rider first must have a soft seat, and in order to become "one with his horse" and "instinctively use the hand and leg aids at the proper time," the rider also must develop feel, or "tact." Once confirmed as soft and tactful, the rider as would-be trainer then must develop a proper pedagogy: he must learn that "dressage training should be controlled gymnastics, not forced exercises"; that "the first task of a trainer should be to preserve the mind of the young horse . . . and protect it against mistrust or fear"; and that "one should always follow the principle: Better too little and too slow than too much and too fast!"[61]

Proper riding and training must follow Steinbrecht's now famous "first main principle . . . ride *your horse forward and set it straight*." The reason is simple: "Thrust cannot be regulated if none exists, and the horse cannot learn to move correctly if it does not move. . . . The horse is a harmonious whole in which the individual parts mutually support one another." In Steinbrecht's estimation, Baucher's violation of this basic principle defines Baucherism and damns its practitioners. Perhaps not surprisingly, Steinbrecht denounces Baucher as a "quack" whose system "consists in . . . robbing the horse of its natural power," until the horse, "no longer good for any practical purpose," is good only for the circus. Baucher's vaunted originality lies only in the fact that "no one before him has discredited the noble art in the way he does by working against all its natural principles." Valuable only as a cautionary, "the Baucher method is a complete system for ruining horses."[62]

Steinbrecht's principles, though proposed late in the nineteenth century, represent a juncture of Romantic and Enlightenment thinking. "The fine arts produce true beauty only if they stay within the confines of nature," Steinbrecht writes, so the proper trainer need "listen only to nature in order to learn the basic principles of producing a good horse."

He urges the trainer, specifically, to "diligently observe the young, green horse in the pasture" to appreciate its natural endowments; to shun "artificially construed bits or other devices"; and to heed and apply natural principles and increase "organic forces." His demand for "the scientifically educated and experienced trainer" who knows and applies "correct, scientific principles" follows from (rather than deviates from) adherence to the laws of nature, including the law of statics that Steinbrecht brought forward from Weyrother and Seeger. Steinbrecht, in short, has a Romantic vision of nature, but clearly an Enlightenment concept of nature as a force not to be celebrated for its wildness but rather to be ordered by art and science.[63]

Baucher had advertised his "new method" to the military as one that enabled breaking and training cavalry horses in record time, a claim seconded by the British cavalry officer Lewis Edward Nolan, in his Baucherist treatise, *The Training of Cavalry Remount Horses, a New System* (1852). For Steinbrecht, the claim is moot: Baucherized horses are ruined horses, no matter the intended purpose nor the time needed to achieve it. With correct preparation of cavalry horses (and riders) as arguably *his* primary goal, Steinbrecht extended the German and Austrian tradition of employing *manège* work to the degree appropriate for military riding. Thus, he carefully distinguishes degree of contact, collection, and self-carriage needed by military horses (as opposed to *haute école* horses, on one side, and racehorses, on the other), and how to achieve it. As Plinzner restates Steinbrecht's thinking, "Correct, truly gymnastic dressage training can never unfavorably influence the performance of a horse cross-country."[64]

Steinbrecht, in the end, exerted profound and lasting influence on both *manège* and military riding—at least more, with respect to the latter, than did Baucher. Steinbrecht's *Gymnasium* directly influenced *H. Dv. 12: Army Riding Regulation 12* (1937), a manual for German cavalry that continues to play a pivotal role in German dressage today. In the preface to a recent English edition of *H. Dv. 12*, Richard F. Williams recounts that German cavalry regulations dating to the eighteenth century were consolidated in 1882, revised in 1912 and 1926, and published in their twelfth version in 1937 (following by only two years, though perhaps coincidentally, the revised fourth edition of the *Gymnasium*). Significantly, "The 1912 edition introduced Riding Provisions that were substantially the work of Steinbrecht." As Christoph Hess adds, "After World War II, the teachings of the *H. Dv. 12* served as the basis for the 'Principles of Riding and Driving' of the German National Equestrian Federation (FN)." Though the FN Principles handbook has been updated often, Hess adds, the "actual core of the content" remains that of *H. Dv. 12*.[65]

De Bragança, to sum up, has contrasted exactitude as "the primary occupation of the German School" of dressage with beauty as "the primary preoccupation of the French school." He goes on to say: "The German School, which has always applied the teaching of the Old French School with a view to obtaining great precision in the movements, was in open war with the Baucherism that was presenting a new method to explore every horse's possibilities."[66] The contrast rings true with respect to dressage as *manège* schooling, as opposed to dressage as "training" in general. Cavalry, however, was reaching its highwater mark in the nineteenth century. European nations raised and maintained huge establishments, requiring the training of untold thousands of horses and men in short order. Schools of dressage and equitation had to find highly efficient and effective methods to feed the insatiable maw of the military. They had to create mechanistic processes for producing mounts and riders, in effect, who themselves neither were machines nor should become machines. German exactitude and French aesthetics in dressage in the nineteenth century must be viewed through that lens.

Outdoor Baucherists

While it is unlikely that Baucher, as James Fillis testified, "never rode outside," the enclosed school, indoors or out, was his pedagogical home and where he made his formidable reputation. General L'Hotte, student of Baucher *and* d'Aure, theoretician and practitioner of a melded *manège* and cross-country equitation, obviously did ride outside on a regular basis, as did, if perhaps to a lesser extent, the English-born French master Fillis, a "follower" of Baucher, a consummate *manège* rider, and a willing and very capable outdoor rider. The cavalry officer General Faverot de Kerbrech, devotee of Baucher and L'Hotte, traveled further than Fillis from *manège* to field in his writing and work, and Captain Étienne Beudant, as the title of his primary work, *Horse Training, Outdoor and High School*, clearly indicates, made that travel his mission.[67]

Forward, always forward

Général Alexis-François L'Hotte (1825–1904), pupil and confidant of both Baucher and d'Aure, exerted powerful institutional and intellectual influence on French military riding in the late nineteenth century. Appointed *écuyer en chef* of the École de Cavalerie at Saumur in 1866, commandant of Saumur in 1875, and member and president of the Comité de Cavalerie in 1886, L'Hotte also published two important books: *Questions équestres* (1895) and a posthumous memoir, *Un Officier de cavalerie—Souvenirs du general L'Hotte* (1906).[68] In those publications L'Hotte initiated the serious efforts to accommodate, if not exactly to reconcile, the principles of Baucher and d'Aure in a hybrid system of *manège* and cross-country equitation suitable for modern cavalry. "These two types of equitation," L'Hotte writes, inform "the questions I put forth inspired by my two masters"— that is to say, they inform his *Questions équestres*.[69]

Wasting no time there in establishing his lineage, L'Hotte invokes "my two illustrious teachers, Baucher and d'Aure," in the opening sentence of *Questions équestres* and often refers to "these two great rivals" as his "two masters." Exemplifying the *manège* and cross-country styles of equitation considered antithetical when L'Hotte entered Saumur as an officer-pupil in 1845, Baucher and d'Aure also represent positions that L'Hotte when a commandant had to negotiate politically as well as conceptually. Obliged in his official capacity to affirm d'Aure's method (the basis for the cavalry's Regulations of 1876), and also advised for political reasons to play down his debts to the institutionally rejected Baucher, L'Hotte became free in his subsequent writing to acknowledge those debts, as well as his departures from Baucher, while still honoring d'Aure. Thus, in *Questions équestres*, he accords "d'Aure, this greatly admired centaur, and Baucher, this unparalleled and fertile innovator," equal status as "the two great examples of French equitation."[70]

Like his predecessors, L'Hotte places "simple" elements of equitation in complex and subtle relationships. His goals "can be expressed in three words: *calm, forward, straight*," and his basic concept in only a few more: "And the straight position is the touchstone to lightness," with flexibility mediating between them. To develop a horse that "executes every movement and in all situations, and shows the desire to go forward," the *écuyer* first must achieve "a rigorously straight horse, straight from head to haunches." Why? A horse's submission and forwardness "[rest] upon the flexibility of his joints . . . obtained jointly with impulsion," and the rider obtains flexibility through "the *ramener* and the *rassembler*, whose perfection depends upon a common source, an *ur*-source . . . which is *the straight horse*." Through that chain, in short, the "straight position" leads to lightness, "the perfect obedience of the horse at the slightest indications of the hand or the heels of the rider."[71]

Since lightness "emanates . . . from the accuracy of the actions of the rider," L'Hotte devotes almost as much attention to the rider as he does to the horse. A school rider must avoid "disgraceful movements," such as twisting his body or shifting his buttocks in "successive lead changes"; both the school and cross-country rider must learn instead to distribute their weight to modify the horse's equilibrium, a skill in which "d'Aure excelled." The educated rider must ensure that "movements of hands and legs must be secret and remain invisible to the eye," and that "everything that draws attention to his person must be avoided. . . . It is the horse who is executing the movements, the rider must merely try to be in harmony with him." Put differently, "every equestrian activity must include . . . 'perfect timing and good measure,' that is, *equestrian tact*." Having perfected lightness, the school rider then can achieve the even higher goal of the quest: "*A horse who moves and handles himself as though acting on his own*," and who, at the right moments, acts *in fact* on his own.[72]

Though L'Hotte emphasizes "classical or *savante* equitation" in *Questions équestres*, he connects it to cross-country and military riding (and also expressly disconnects it from circus riding).[73] With respect to conformation, for example, military horses and cross-country horses should have comparable though different proportions; and with respect to *manège* training, "Its importance lies in the disposition of the haunches which must also be a part of cross-country equitation. Cross-country equitation, it is true, does not lay claim to controlling the strength of the horse which characterizes Classical or *savante* equitation, whose trait is the purity of harmony. But one must still dominate the forces of the horse to the extent that this is necessary for his ordinary use." Overall, troop horses and riders should be able to execute "straight lines and turns . . . with precision and regularity at the three gaits," and they will improve in health and strength if trained outdoors as well as in the *manège*.[74]

British by birth but long resident in France, and, for a period, *écuyer en chef* of the Russian Cavalry School, L'Hotte's contemporary James Fillis (1834–1913) studied under François Caron, who had studied under Baucher, and earned a reputation as one of the most skilled and accomplished riders in fin de siècle Europe. Fillis published two books, both originally in French. *Principes de dressage et d'équitation* (1890), the more influential of the two, was "edited" by the eminent statesman Georges Clemenceau and translated by the distinguished horseman M. Horace Hayes.[75] The book unfolds in eleven chapters of widely varying length and detail. Two have special pertinence to our concerns—"Commentaries on Baucher" and "The Army Horse"—as do close, independent, and unexpected alignments between Fillis's and Steinbrecht's thinking.

Published in English for decades as *Breaking and Riding*, Fillis's treatise, its title more literally translated as "Principles of Dressage and Equitation," is a model of lucidity. Fillis sets the tone in his admirably terse preface: he does "not presume to discuss scientific subjects," because he is "simply a horseman" who can draw on "sixty years" of experience. He derives from that experience the "fundamental principle" that a horse must be "correctly balanced and light in forward movements and propulsion in order that the rider may obtain the most powerful effects with the least exertion." And he encapsulates the method based on that principle as consisting in "distribution of weight by the height of the neck bent at the poll and not at the withers; propulsion by means of the hocks being brought under the body; and lightness by the loosening of the lower jaw"—adding the jaunty comment, "When we know this, we know everything, and we know nothing." We know, that is, the "universal" principles, but not yet the proper application of them that we presumably will learn from his book.[76]

Attuned equally to horses and riders, Fillis is studiously unsentimental, believing horses "incapable of affection for man" and indifferent "towards those who tend them or ride them"; though horses possess "particularly acute" memory, moreover, they have "but little intelligence" and "cannot reason." He is unconditional on temperament: the "supreme quality in every horse is impetuosity," and "every horse which is not hot is fit only to be put between the shafts." And he is unreserved on breeding: "I unhesitatingly put Thoroughbreds above all others, whether for hacking or for high school riding. They are pre-eminently the best for all kinds of work" . . .

except military.[77] While he clearly follows Baucher, on the one hand, in demanding total submission—"my great point is to be master. . . . The object of training is the destruction of the free will of the horse"—he also disparages, on the other hand, the "new school" that produces only "machines and automatons."[78]

With respect to riders, Fillis shares with Steinbrecht the belief that only a proper rider can be an effective trainer and that "the supreme quality of a rider" is equestrian tact. Fillis's definition of tact maps closely onto Steinbrecht's in prerequisite ("flexibility of the limbs and a good seat"); effectiveness ("send[ing] to the hind quarters only the amount of force necessary to maintain equilibrium with a maximum of propulsion"); and manifestation ("being able to feel that the horse is straight"). Fillis also shares with Steinbrecht a commitment, clearly *not* Baucherist, to *forward movement.* Since a horse not "always ready to go forward . . . is useless," stationary work in training, not only "a deplorable waste of time," impedes the obtaining of "forward movement . . . that has to be obtained at any price." Thus, Fillis asserts, once mounted, he never asks his horse "anything except when he is advancing," a d'Aurist axiom that explains Fillis's adoption of the d'Aurist motto: "forward, always forward, and again forward!"[79]

This all comes back to Baucher, about whom Fillis showed more ambivalence than did Steinbrecht. Fillis opens his "Commentaries on Baucher" by calling him "certainly the greatest and most clever high-school rider we have ever had," adding: "As regards myself, I claim to be a follower of Baucher. . . . Without [him], I would not know as much as I do of riding." In more than a score of previous references to Baucher, however, almost all critical, Fillis had emphasized a single point: "Finally, the work of making the horse to go forward constitutes the great difference between my system of equitation and that of Baucher." And, in the process, he also makes a politic, if suspect, move: Baucher's "marvelous equestrian tact remedied every deficiency. Where his theory was false, his hands and legs [rectified] the error of his doctrine." Baucher, unfortunately, "could not put his tact into his books." Fillis, in a word, proffers left-handed compliments where Steinbrecht hurls invective.[80]

Ultimately, Fillis represented his system—indeed, *had* to represent it—as advantageous to the successful waging of war. "Late wars," he notes, "have proven that cavalry are required to play a decisive part in military operations"; cavalry require trained and obedient horses; and other training methods waste time, effort, and money, the last still "the chief sinew of war." The remount depot, therefore, "ought to be a true breaking school," teaching equestrian equilibrium to soldiers and preparing horses through a proper system of dressage. "It is an interesting anomaly," Felton writes, "that whereas both Baucher and Fillis greatly influenced the thinking of most civilian riders and authors, Baucher exerted relatively little influence and Fillis practically no influence at all on the development of equitation at the great French School at Saumur." Moreover, Chenevix-Trench adds regarding England, since "the average British cavalry officer was an amateur . . . the result of attempts to school British cavalry-horses on the methods of Baucher and Fillis was uniformly unhappy."[81]

Indoor and out

Baucher, metaphorically speaking, sired not only L'Hotte and Fillis but also François Faverot de Kerbrech (1837–1905), just as Faverot de Kerbrech (henceforth, Faverot), in turn, "sired" Étienne Beudant (1863–1949). The title page of Faverot's principal work, *Dressage Méthodique du Cheval de Selle* (1891), carries the subtitle *D'Après les Derniers Enseignements de F. Baucher, Recueillis par Un de Ses Élèves* (*Methodical Dressage of the Riding Horse, From the Last Teachings of F. Baucher, As Recalled by One of His Students*); and the author's dedicatory letter in Beudant's main work, *Extérieur et Haute École* (*Horse Training: Outdoors and High School*) (1923) acknowledges "that all I have done has been to practice the equitation doctrines of General Faverot de Kerbrech."[82]

In *Methodical Dressage of the Saddle Horse*, Faverot published a treatise as methodical as the dressage it advocates. An introductory section, *Definitions, Goal of Dressage, and General Principles*, offers definitions of four key terms—action, position, movement, and balance; offers a statement of the fundamental goal—"any saddle horse must be rendered easy and agreeable to ride, regular in his gaits (paces), docile, willing, and as brilliant as his conformation allows"—with five corollary goals; and offers five corresponding general principles that range from the *ramener* to the *rassembler*.[83] The body of the text, *Progression of Dressage*, comprises four lengthy sections, each comprising multiple chapters, that unfold and detail a systematic program of equine training, the first two sections on "preparation," the third on "*assembler*" (putting it together), and the fourth on "confirming the horse."

That program, Faverot made clear, is Baucher's. In addition to his book's subtitular reference, the first edition featured an illustration of Baucher on its wrappers and an engraving of "Baucher on Partisan" as its frontispiece. More important, an unsigned foreword (at least hypothetically attributable to Faverot) notes that Baucher, retiring from public riding following his serious accident, never integrated in writing or teaching the principles that had evolved over many decades as his First and Second Methods, but instead "preferred to pose [and show how to apply] general principles . . . to teachers, trained directly at his school." One of those teachers—an intimate who "saw Baucher work, and has himself ridden the last horses trained by that incomparable écuyer"—Faverot intends to preserve "the detailed lessons of this learned master . . . threatened with disappearance without leaving a trace."[84]

Though Faverot's lessons themselves defy brief summary, their basic principles stand out. Following Baucher, Faverot teaches that "*lightness* must be the constant preoccupation of the rider," straightness "the object of constant care," and self-carriage an essential goal; he teaches that application of tact and finesse rivals application of theory or system; and he teaches, to cite two specific Baucherist axioms, one, the importance of legs without hand, hand without legs ("If the leg and the hand act simultaneously, their effects tend to annul each other and produce contractions"), and two, the priority of position over movement ("In a word, give the *position* which should cause the movement, and do not look for the *movement* which must be a consequence of the position")—the reason, as noted earlier, behind Baucher's predilection for work in place and for the walk as a training gait.[85]

Faverot "concludes" that he "wanted to reveal in this study *all* of the procedures of dressage used by Baucher at the end of his life" (that is, in the Second Method) and adds that, in fulfilling that goal, "he has presented the means to push the horse's education all the way to a sort of *ideal* perfection, which to his knowledge only the horses of Baucher and those of General L'Hotte have attained." Those procedures rest on two basic precepts. One, deliberate, calm, and kind training, as distilled in three epigrams, ensures both economy and harmony: a) "Ask often; be content with little; reward generously"; b) "Go very slowly to bring the training along rapidly"; and c) "Demand very little at the beginning. Be content with little." Two, economy of means, invoked in Faverot's final sentence, ensures both efficiency and effectiveness: "The wiser the horse and the more modest the rider in his demands, the more likely the rider can simplify the training and reduce the number of means used."[86]

Though little in *Methodical Dressage* applies specifically to training the outdoor horse, very little in it cannot be applied to that purpose. Work in hand, for example, can hold value for a hunter, likewise *rassembler* for an army horse, and, more obviously, "turns by the pressure of the reins . . . to the military rider who conducts his horse *with only one hand*" (the other wielding saber or lance). The trainer, however, must distinguish what is necessary for a horse "specifically destined to the outdoors," as opposed to one being prepared for "higher equitation"—a distinction that Faverot later explored in a pithy text, *Dressage du cheval de dehors, conseils donnés aux membres de l'Étrier* (*Dressage of the Outdoor Horse, Advice Given to Members of* ["the Stirrup" Riding Club], *Recalled by one of* [Faverot's] *students, General George de Lagarenne*) (1907).[87]

Étienne Beudant's elegant treatise, *Horse Training: Outdoors and High School*, shifts Faverot's emphasis from school to outdoor training. More fluid than *Methodical Dressage* in structure and style, *Outdoors and High School* opens with Beudant's fully developed thoughts on horse training and methods and on "hands without legs, legs without hands" (accorded pride of place here), and then offers lucid definitions of thirteen terms—some in one or two sentences, others in one or two pages, all more expressive than technical: "Tact: The genius of equitation—the feel of the horse."[88] After wending through ample discussions of topics including lightness, progression of dressage, obedience to the spur, *ramener* and *rassembler*, work at the canter, jumping, and so on, the book closes with results of training, comprising instructive précis on the training of many of Beudant's horses (notably, the gifted Robersart II and the timid Mimoun).

Beudant spins an intricate web of influences. "All precepts [for training methods] come from the teachings of Baucher and Comte d'Aure, the two great masters of the modern art of equitation." Though their principles are the same, in Beudant's view, their emphases differ: lightness for Baucher, the great *manège* rider, and impulsion for d'Aure, the great outdoor rider. General L'Hotte, *"the greatest equestrian genius that ever lived,"* balanced and melded them. "A pupil worthy of his two great teachers, Comte d'Aure and Baucher, [L'Hotte] was the foremost horseman of his day." Any success that Beudant has had "in the training of my horses," however, also pays "homage" to Faverot (invoked always as "General Ferverot de Kerbrech"), whose *Methodical Dressage* has such simplicity and clarity that "any one can understand this most practical of all methods [Baucherism]. This is why we owe so much to [its author]." Thus, Beudant concludes his patrimony, "Two masters have been my guides, [Faverot] and General L'Hotte."[89]

More attuned to philosophy than to technique (though excelling at both as rider and as writer), Beudant also develops broad themes rather than expounds narrow principles. The rider's task and responsibility are "*to observe* and *to reflect*"—to observe the horse at liberty and to reflect on his own, not his mount's, errors—and to communicate; but when communication fails, "The fault is nearly always with the rider." Intelligent and willing, the horse "habitually responds logically to demands made upon him," and "almost always . . . tries his best to respond to orders which in most instances are poorly expressed." For these reasons, "The rider must reduce *his actions to the very minimum* and leave the horse *the greatest possible freedom in his.*" This applies particularly outdoors, where the horse's instinct "serves him infinitely better than he could be served by even the most skilled rider," and where the horse should be "as supple and brilliant under saddle as at liberty" (the final word appears often).[90]

Beudant gives a modern twist to this fundamentally Romantic view. He derides "the emptiness of theories and scientific formulae for the training of a riding horse," and he excoriates "the experts of the different schools of training" and their "learned demonstrations frequently backed up by figures and calculations in mechanics." Proper dressage, instead, relies on tact and passion: "The understanding of a horseman comes from his heart, not from mathematical calculations. Why then try to make of his noble comrade a mere automaton?" Such calculations and the theories based on them "do not teach us how to take our horse WHERE WE WISH, WHEN WE WISH, AS WE WISH." They only reflect and further encourage "the pretensions of those who . . . believe that they can manage and care for a horse as a mechanic guides and repairs a motor car." In a word, abstruse theories that communicate poorly with riders produce inept riders who communicate poorly with horses.[91]

It remained for Beudant, like L'Hotte, to bring Baucher and d'Aure into alignment. "The greatest pleasure to be got from horses is to gallop a well-trained one over varied country, dotted here and there with jumps,"

Beudant writes, "and I cannot understand why anyone would try to obtain the more or less stiff and fantastic airs of the school before making an agreeable and pleasant outdoor horse." The answer lies in the phrase "well trained" (or, elsewhere, "perfectly trained"): *haute école* airs and gymnastics may be "entirely unnecessary for ordinary training, [but they] are the best means by which to arrive at the perfection of that training. . . . The *haute école* and equitation of fantasy are at once both a refinement *de luxe* and a practical means of arriving at perfection." In short, Beudant makes a distinction between training and preparation for training that dates to de Pluvinel: "The *haute école* balances and perfects [the horse] for all service; gives him carriage."[92] The military rider then can use that carriage to perfect an outdoor horse.

In addition to literary finesse, *Outdoors and High School* has a marked elegiac quality. Beudant opens its dedicatory letter by recalling that the addressee, General Juinot-Gambetta, had promised at the beginning of the Great War "to send me to the great battle-front," but that a subsequent, and severe, riding accident had precluded that service. "Pain, only pain which never quits me," Beudant recalls his emotion, "mitigates the bitterness of my inaction. All is over for me—no more the joy of equitation." Extending the theme in the opening of his text proper—"I studied solely the training of the riding horse," a horse soldier "unable to march to victory, as did real horsemen"—he concludes it chapters later: "Now I ride Mimoun on solitary quiet rides . . . dreaming, sometimes with sadness, of hopes blasted and gone awry. Once I anticipated ending my days with the Cavalry among horsemen and beautiful horses." A common trope in Great War literature, injury and inability to perform mean loss of futurity, potency, and love.[93]

Though the principles for outdoor riding and jumping advanced by Kerbrech and Beudant, finally, did not fall into oblivion, the revolutionary theory and practice concurrently developed by the Italian cavalry officer Federico Caprilli (1868–1907) in 1904–1906, just prior to his untimely death, would eclipse them almost fully in the twentieth century.[94] Initially defined, interpreted, and widely disseminated as the "Italian method" by Caprilli's indefatigable apostle, Piero Santini (1881–1960), in the latter's trilogy *Riding Reflections* (1932), *The Forward Impulse* (1936), and *Learning to Ride* (1941), Caprilli's system—subsequently known as "forward riding" or the "forward seat"—would come to define outdoor equitation, military and civilian, and to dominate international show jumping, not only in Italy, but also in the United States and elsewhere, largely through the writing and teaching of eminent second-generation military disciples such as Vladimir Littauer and Harry Chamberlin. But that is a topic for another occasion.[95]

In brief summary, then, from the sixteenth to the eighteenth centuries, master horsemen guided the evolution of dressage and equitation from *haute école* training as the acme of artistic expression with degrees of military application to a still advanced but more elemental form of *manège* training designed specifically for mounted warfare. Over the course of the eighteenth and nineteenth centuries, horsemen guided that evolution through its next stage: combined *manège* and cross-country training, the latter derived mainly from field sports, as the ideal preparation for military riding. Warfare in general also evolved concurrently over those four centuries. Indeed, horsemanship and mounted warfare evolved in tandem, exerting reciprocal pressures and influence. That dual evolution determined the history of cavalry, especially light cavalry, as we know it.

3

Light-Horse, Dragoons, and Others

You cannot improvise cavalry, nor can you make cavalry by putting a man on a horse and calling him a cavalryman.

Charles Sydney Goldman
With General French and the Cavalry in South Africa (1902)

Following the chivalric era, mounted arms evolved over the late seventeenth and early eighteenth centuries into the two basic forms of heavy and light cavalry. In the mid-eighteenth century, Frederick the Great, king of Prussia, and General Friedrich Wilhelm von Seydlitz, his brilliant cavalry general, next brought light cavalry—what the English called *light-horse*—to a high degree of refinement in equitation and tactics. Progressing through the late eighteenth century, light-horse reached a high point in the Napoleonic Wars of 1796–1815 that "marked a revolution in the politics of Europe and in the history of warfare," to quote Louis A. DiMarco in *War Horse* (2008). Cavalry continued to metamorphose as a dominant military arm over the long nineteenth century. As Stephen Badsey has noted in *Doctrine and Reform in the British Cavalry 1880–1918* (2008), "The extent to which European land warfare was transformed between the early nineteenth century and early twentieth century almost cannot be overstated." The cavalry arm and its horse soldiers played a central role in that transformation.[1]

The role and tactics of cavalry over the long nineteenth century, specifically, transformed from *shock action*, ridden in close formation on battlefields by soldiers mounted on large horses and armed primarily with steel weapons, to *fire action*, ridden in loose (or no) formation in often far-flung detached units by soldiers mounted on lighter and quicker horses and equipped primarily with firearms. The last stages in the long evolution from *manège* training of military horse and rider to a combination of *manège* and cross-country training—the evolution traced in the preceding two chapters—corresponded to that transformation in cavalry role and tactics. The horse soldiers who wrote treatises, manuals, and other works on cavalry in this long century, virtually all of them officers, typically focused on the arm itself—its history, mission, organization, strategy, and tactics. Horse soldiers of rank, however, often were accomplished horsemen, and those who wrote books on cavalry nearly always attended, to one degree or other, to horses and horsemanship. If fighting was the raison d'être of cavalry, horses were the indispensable resource and horsemanship the means for deploying it.

Cavalry Terms

Books on cavalry published in the long nineteenth century collectively covered untold topics and subtopics from the abstruse to the quotidian, from the metaphysics of "cavalry spirit" to the mechanics of carbines and revolvers.[2] Those topics included a number of terms and concepts with direct or indirect bearing on military horses and horsemanship:

Military *doctrine*, or "the prescriptive setting out of the courses of action that armed forces should follow," in Badsey's words, determined strategy and tactics, "The two great divisions of the Art of War," as Arthur L. Wagner wrote in 1895: *strategy*, or "the art of moving an army in the theater of operations . . . placing it in such a position . . . as to increase the probability of victory, increase the consequences of victory, or lessen the consequences of defeat"; and *tactics*, or "the art of disposing and maneuvering troops on the field of battle." Tactics entailed evolutions, or synchronized movements of men and horses in formation, and maneuvers, or combinations of evolutions in battle. They all served the fundamental object of war: in General Friedrich von Bernhardi's words, *"to impose our will on that of the enemy."*[3]

Cavalry comprised heavy or line cavalry, light cavalry or light-horse, and dragoons, distinguished by type of horse and man, armament, and, especially, role. *Heavy cavalry*, who mainly charged infantry en masse in battle, were superseded by light cavalry, who then conducted all actions, including detached duties of reconnoitering the enemy and protecting (or, "covering") their own troops. *Light cavalry* might be "divisional" (attached to an infantry division for protection) or "independent" (unattached and used for reconnaissance).[4] *Dragoons*, originally "mounted foot soldiers [who] always fought on foot" and used horses simply for rapid transit, as Friedrich Wilhelm Bismarck wrote, evolved into what Lewis Edward Nolan called "a sort of hybrid corps [of] horse and foot soldiers." By the end of the nineteenth century, Wagner could write, "The dragoon is essentially the cavalryman of the present day . . . equally capable of mounted action with the saber and dismounted action with the carbine," and so true cavalry, as opposed to mounted infantry, who were able to fight only dismounted with rifles.[5]

Mobility conferred by the horse was the foundational property of cavalry; everything from doctrine to tactics derived from it. "From time immemorial," Erskine Childers wrote in *War and the Arme Blanche* (1910), "the superior mobility derived from the horse . . . has given to Cavalry . . . all the special functions which distinguish it from Infantry." Mobility had corollaries: generally speaking, speed referred to velocity of mounted equines; rapidity to dispatch in employing velocity or deploying units; quickness to phenomena ranging from the physical agility of troopers to the mental agility of officers; and promptness simply to arriving on time, no small matter given the logistics of moving masses of horses and troops. Horse artillery, "specially designed for service with cavalry," Wagner notes, also had mobility as "its essential characteristic." While field artillery accompanying cavalry impeded its mobility, horse artillery—its versatile horses pulling lighter guns while being ridden by gunners—kept pace with cavalry on both road and battlefield. "Only a madman or an absolute ignoramus," M. F. Rimington wrote in *Our Cavalry* (1912), "would willingly dispense with horse artillery."[6]

The intangible counterpart to mobility, "cavalry spirit," a term "so often used that it was hardly felt to need definition," had emerged historically in tandem with shock action to the degree that cavalrymen thereafter often considered them inextricable. Usually discussed with morale and esprit de corps, "cavalry spirit" was associated with offensive, as opposed to defensive, zeal and action, and represented, in Colonel Monsenergue's words, "the fearlessness of responsibility, energy and activity, with a keen desire always to assume the offensive, and attack or counter-attack the enemy." What constituted "cavalry spirit" was hotly contested in the late nineteenth and early twentieth centuries in the "*arme blanche* controversy" that pitted advocates for mounted action with cold steel, who believed that dismounted fire action either lacked or corrupted the "true cavalry spirit," against advocates of dismounted fire action, who dismissed that belief as reactionary nonsense. In their view, cavalry based on dismounted fire action not only retained the offensive mobility of cavalry but also gained the defensive stability of infantry: it simply *redefined* "cavalry spirit."[7]

A cavalry *charge* against infantry converted mass and speed into *shock*, a formidable combination of physical assault and psychological terror. Crashing through the enemy's lines "by sheer momentum and weight of numbers," dispatching him with cold steel, the charge, as noted above, was synonymous with cavalry and "cavalry spirit," and certainly was not without danger. Robert Hinde, in 1778, called the cavalry charge "undoubtedly the most dangerous service that cavalry can be employed in"; and Bismarck, in 1819, saw it as "very uncertain, and the least unexpected resistance often causes it to fail." Cavalry charges against cavalry, also at speed and in close intervals, were neither easier nor less risky. As Nolan wrote in *Cavalry: Its History and Tactics* in 1853, "Cavalry seldom meet each other in a charge executed at speed; the one party generally turns before joining issue with the enemy. . . . Every cavalry soldier approaching another at speed must feel that if they come in contact at that pace they both go down, and probably break every limb in their bodies."[8]

Cavalry eventually shifted its primary role from charging in battle to reconnoitering the enemy and covering its own army's main body in the field, often referred to as the services of *information* and *security*. No one disputed the fact, as Bernhardi wrote, that "cavalry has always to face the double task of simultaneously reconnoitring and screening [or covering]; and it will often have to decide on which of these activities it has to lay the greatest stress." Writers debated, however, what constituted each duty and the relationship between them; how best to execute them individually; whether or not the same detachment could execute them at the same time; and whether or not all men and horses were capable of performing them. Rimington offered a nice analogy in support of separating them: "[Cavalry] can not be relied upon to perform efficiently the duties of policeman and detective at one and the same time. The duty of the latter would carry the former away from his beat."[9]

Arme blanche, or *steel weapons*, comprised the lance and the saber or sword. Often called the "queen of weapons," the lance produced a chilling battle-field effect, but it also had limited application, demanded considerable strength and skill, and so, as Jon Coulston notes, "went in and out of mili-tary fashion in nineteenth century armies." Not so the saber, "indisputably the most formidable and essentially useful weapon of cavalry," as L. A. Neville proclaimed in 1796; the cavalry weapon "par excellence," as George Denison and Wagner wrote nearly a century later; the "chief weapon," in Carl von Schmidt's view, after the "horse, rapidity, and mobility."[10] The lit-erature referred relentlessly to men or troops being "cut to pieces" by sabers, one action in a cavalry duel, for example, being to slice off either the bridle or the sword hand of one's adversary—each achieving the same result. Cav-alrymen debated the comparative merits of using the saber to cut (slash) or thrust (point), resulting in ongoing redesign of the saber and ongoing revi-sion of prescribed exercises in swordsmanship; and they repeatedly invoked the "moral as well as physical effect . . . produced by a body of men cutting and slashing at all around them."[11]

What the saber was to a mounted charge, pursuit, or retreat, so the *carbine*, or short-barreled rifle, was to dismounted cavalry action, so that "by 1885," DiMarco writes, "most European cavalry carried both a sword [saber] and an effective breechloading carbine." While advances in infantry firearms reduced the effectiveness of cavalry on the battlefield, advances in cavalry firearms, namely the carbine, increased the arm's effectiveness in information and security primarily by enabling cavalrymen to fight dismounted.[12] Despite that benefit, the cavalry carbine had two drawbacks: one, it was not effective in mounted action; and two, it was not so effective as an infantry rifle.[13] Although neither drawback could be eliminated, both could be mitigated. In addition to the carbine, the revolver could supplement or replace the saber and produce an excellent weapon for mounted action at close range; and the infantry rifle in the hands of mounted riflemen, or mounted infantry, could provide a force for dismounted action at long range. After the turn of the cen-tury, finally, military writers in several countries promoted incorporation of machine gun units by cavalry.[14]

Treatises on leadership and histories of famous leaders typically described the traits, skills, experience, and knowledge of an ideal *cavalry officer*, frequently in the language of exceptionalism. "Energy, indefatigable, indomitable energy—is perfectly invaluable in a general, and a cavalry general requires it most of all," as Denison wrote in *A History of Cavalry* (1877, 1913).[15] Manuals and regulations, serving a different purpose, prescribed the duties of an actual officer, their proper execution, and the applicable standards for evaluation. A cavalry officer should be inspirational and exemplary; prudent but courageous; cool but impetuous; resolute but improvisational; and decisive, quick thinking, and imaginative. Such an officer should have practical experience in warfare, but, in any case, also must study both the history of warfare and the lessons of recent wars.[16] A cavalry officer, finally, must be an accomplished horseman and horsemaster with "a thorough knowledge of horse management, training, and equitation" and must be "continually on the alert, looking out for the welfare of his horses."[17]

Horsemanship, Bernhardi wrote, is "absolutely the bed-rock of all Cavalry performances." While all horse soldiers were nominally horsemen, however, they also varied greatly in knowledge and skill. British cavalry officers, for example, typically entered service as accomplished riders by virtue of social standing and foxhunting experience, while recruits, by contrast, often had little or no equestrian experience. Officers were expected to turn them into reasonably capable horsemen, sufficiently competent in both dressage and equitation to keep themselves and their fellow recruits alive. Individual men and horses constituted collective battle formations, and only well-trained men and horses could form, hold, and fight in them. Nolan's caution, "Everyman may be taught to ride, but it is not every man that will make a good rider," therefore, had weight: one inept rider on one uncontrolled horse could throw a tactical maneuver into disorder and confusion. "The power of cavalry in the field," Nolan explained, "depends upon the individual efficiency of the horsemen composing it; infinite care should, therefore, be bestowed on their training and teaching."[18]

Cavalry depended on mobility, mobility depended on horses, and horses depended on capable *horsemasters* and horse management as much as on horsemen.[19] For mounted cavalry to remain reliably mounted, officers and men had to provide horses love and overall care; guarantee their fitness and condition; safeguard their health and nutrition; oversee their proper shoeing and tack; ensure their safe transport and shelter; and much else.[20] So much else, indeed, that training manuals and guides almost always treated the care of horses in camp, on the march, and in action at length and in detail. "Cavalry is always dependent upon the condition of the horses," Denison wrote in 1877. "If they are not in an efficient state, if their shoes are not carefully looked to and sore backs guarded against, they are destroyed, and the force becomes worthless." Thus, Rimington advised young officers in 1912, "From the day he joins, no opportunity should be lost of teaching the recruit that amongst his first duties is to love, honour, and have a pride in his horse." The arm's singular and expensive asset, the horse was to be protected with vigilance at all times and at all costs.[21]

When Bismarck observed that the "strength [of cavalry] is never reckoned, as in infantry, by the number of men, but by the number of horses," he was illustrating the broader point that the cavalry arm, in effect, attached more importance to the horse than to the man. Cavalry had to train horses and men in peace to have them ready for war; and officers had to invest as much time and effort in their horses as in their men, if not more. In addition to proper breeding and training, *cavalry horses* had to display mobility, speed, stamina, and obedience, and had to possess boldness and mettle, since the horse was the cavalryman's principal weapon, notably in its weight and speed in shock action, as well as his means of transport. To cite Nolan's encomium, "Saddles will be emptied, horses killed and wounded, but no horse, unless he is shot through the brain, or has his legs broken, will fall; though stricken to the death, he will struggle through the charge." In the end, in sum, the cavalry horse not only provided mobility, but also proved a formidable, fast, and very tough animal, one difficult to evade, and even more difficult to take down.[22]

The writers who drove the conversation on those topics for well over a century, whatever their motives, together shaped an argument for horse cavalry based on the high rate of return on investment, literally and figuratively. Over time, moreover, they transformed the defensive rhetorical strategy that cavalry was not "a thing of the past" into the offensive rhetorical strategy that the arm's "sphere of action" had increased and that cavalry had, "if anything, a greater future before it." That future, like the past, would depend on mobility, horses, and horse soldiers.

Enlightenment on Horseback

Jacques-Louis David's highly idealized equestrian portrait, *Bonaparte Crossing the Alps at the Saint-Bernard* (1801), depicts the young general mounted on his magnificent charger in the foreground, leading troops in the background, his horse *forcené*, his right arm gesturing both ahead to the summit and forward to the future: Napoleon is "calm, mounted on a fiery steed," just as he had instructed David to depict him. The dynamic between calm and fiery also informs David's rendering of Napoleon's expression. Facing the viewer, at once intelligent and resolute, Napoleon is equally at home in rational thought and in passionate action: an emblem of the "Enlightenment on horseback."[23]

Equitation

Henry Herbert, 10th Earl of Pembroke (1734–1794) served as an officer in the British cavalry from 1752 until his death in 1794, having risen to the rank of general in 1782. He also wrote the influential treatise, *A Method of Breaking Horses, and Teaching Soldiers to Ride* (1761 and 1762), revised and retitled, *Military Equitation; or, A Method of Breaking Horses, and Teaching Soldiers to Ride* (1778 and 1793), a work that addressed "the wretched system of Horsemanship, that at present prevails in the ARMY." As Pembroke advises in his dedication to the king, the system leaves men and horses unprepared "for want of proper instructions and intelligence in this Art." Written for "the use of the Cavalry," *Military Equitation* outlines a new system, with a detailed program, for the training of military horse and rider: it comprises principles and practices, with the accent on the former. As a work of advocacy for the resources needed to enhance the system of horsemanship, it speaks to politicians and staff officers; as a work of pedagogy designed to improve the teaching of horsemanship, it speaks to officers and their riding masters.[24]

Pembroke is unsparing in his criticism of current military horsemanship and in his contempt for self-serving riding masters, incompetent farriers, and dishonest grooms: "There is a great deal of good sense in Xenophon's method of forming horses for war," he writes, "after him, horsemanship was buried for ages, or rather brutalised, which is still too much the case." He spares soldiers and horses from reproach, however, because their faults result from ignorance and not from lack of virtue:

soldiers and horses are educable and their faults correctible. Abhorring extremes—"a coward and a madman make alike bad riders [who spoil horses], though in very different ways"—he urges reason and moderation, patience and gentleness, and simplicity of means: "Every thing in horsemanship must be effected by degrees, and with delicacy, but at the same time with spirit and resolution." Though Pembroke advises that "light cavalry [and] their horses should be very ready and expert in leaping over ditches, hedges, gates, &c," finally, he rigorously advocates for *manège* training as critical for precise battlefield maneuvers.[25]

Pembroke's main point about such training is clear: scale, urgency, and lack of resources constrain the proper training of cavalrymen and horses, particularly with respect to the teaching of refined *manége* techniques. Thus, he offers his method as "not only the best method, (if I may say, the only right one) but also the easiest and the shortest," because it recognizes that the perfect unity of rider and horse sought and prescribed in "the practice and teaching of great masters [is a] perfection . . . not to be expected in the hurry which can not be avoided in a regimental school, where the numbers are so great." Pembroke's work may emphasize principles more than practice, in other words, but those principles, he insists, are based on what is feasible and practical. As Vladimir Littauer has observed, Pembroke was "a follower of the French school [of dressage who] was practical enough to suggest its considerable simplification for the army."[26]

What Pembroke did for the French school, William Tyndale did for Pembroke. In regimental service from 1794 to 1803, Tyndale (?–1830) wrote two short military tracts: *Instructions for Young Dragoon Officers* (second edition, 1796), "intended to instruct the young officer in the part of his duty required in quarters [and in] the business of the field," and *A Treatise on Military Equitation* (1797). The second invokes Pembroke in each of its first two paragraphs. Tyndale, as he says, will not presume to match Pembroke's "knowledge of the art," but he can perform the service of improving "regimental riding masters" who may know horsemanship through experience but cannot teach it because they lack knowledge of its principles. Pembroke's *Military Equitation* "is the best work of the kind in our language," in other words, but it is "too deep and too scientific for a treatise on Military Equitation, and infinitely beyond the comprehension of those whom it is my wish to teach and reform." Explaining Pembroke's and his own principles through their practical application, he offers a "work so plain and simple" that regimental commanders and their riding masters can use it as a guide to teaching.[27]

Seconding Pembroke's advocacy for *manège* training, Tyndale presses a concern shared by Pembroke and a few British contemporaries and commonly held on the Continent at the time. The British national "method of riding," Tyndale observes, "however well adapted to the common purposes of hunting, &c," offers no preparation for precise riding within "large connected bodies of horsemen," and has led the British military to neglect formal horsemanship. Regimental instructors, in particular, have failed to retrain young recruits in "the method of riding which is necessarily adopted in regiments of cavalry, and which must be observed by every individual" to prevent confusion—even chaos—in close formation. Whether Tyndale was right or not, he was reflecting the connection between foxhunting and cavalry training common in British cavalry treatises in the eighteenth and nineteenth centuries. As we have seen in the preceding chapter, however, when light-horse and open formations became dominant, writers pressed the case *for* foxhunting and field riding as preparation, and for combined cross-country and *manège* riding as ideal military equitation.[28]

In a more wide-ranging treatise, *An Analysis of Horsemanship* (1799, 1805), John Adams (dates unknown) "offers my humble attempt, at analysing the science of *Horsemanship* . . . to facilitate and improve a practice eminently beneficial to mankind at large, and more especially to the army"—the second objective underscored by chapters "On Military Riding" and "Observations and Remarks on Teaching Soldiers to Ride" (the latter added in 1805). Since civilian gentlemen, Adams writes, now must "discipline themselves as soldiers . . . I thought a work of this kind might prove acceptable to the cavalry." By "this kind," he means a work of instruction in *manège* riding and breaking at a time when riders prefer to copy "a hunting-groom, or racing jockey" than to learn "to ride in a school," and when "*manege* riding and jockey-riding" hold one another in "reciprocal contempt," an unproductive state of affairs. "The *manege* being the foundation and groundwork of all good riding," he argues, formal horsemanship offers soldiers, "without becoming adepts in the science," knowledge of principles and a firm seat and quiet hands that reduce falls and produce a stable fighting platform.[29]

Adams's chapter on military riding elaborates those points: "military riding differs from the manege, only, by laying aside all extremes," such as severe collection; mastery of "the science of the hand and aids," however, is required for "precision and correctness" of movement in formation and "strength and firmness of a squadron in a charge"; and a firm seat is necessary for a soldier "to be expert and ready at the sword exercise on horseback."[30] His chapter on teaching military riding rests on the premise, "Teaching to ride is a separate and distinct art from riding"—an art that requires both theoretical and practical knowledge. The military instructor needs both, in effect, to ensure that troopers learn to coordinate a correct seat and true use of the hand; learn not to "attribute their own errors to the horse" but, instead, to acquire the "extreme delicacy and gentleness required in horsemanship"; and, finally, learn to ride with "precision," for "Nothing will contribute more to the steadiness of the ranks than perfecting the men to the greatest exactness in these apparent little things" that together result in a regiment moving "like clockwork."[31]

Though more narrowly focused than Adams's treatise, finally, Strickland Freeman's *The Art of Horsemanship, Altered and Abbreviated, According to the Principles of the Late Sir Sidney Medows* (1806) also contributed to early nineteenth-century thinking on military horsemanship.[32] More a treatise on dressage than on equitation, *The Art of Horsemanship* proceeds through a sequence of technical chapters on breaking and training young horses. Identifying *manège* training and riding as his area of expertise, Freeman acknowledges, however, that "manage and military riding are so much connected, I hope to be excused for the remarks which are here made on military matters." Like Adams, he argues primarily for *manège* dressage and equitation as the best indirect preparation for military horses and riders: he concludes his book with a lesson on *manège* training and riding to be taught out of doors to "a few of those real lovers of the art, who are to be found in almost every regiment." But even ordinary troopers, he adds, need suppled horses and firm seats, because "when pursuing or pursued by an enemy, when galloping over rough and uneven ground—those men would be likely to succeed the best, whose bodies, as well as their horse's, are poised [or, balanced] the best."[33]

. . . *and tactics*

Those prescriptions for military equitation aligned with contemporary theories of mounted action. In the introduction to a modern reissue of Emanuel von Warnery's influential tract, *Remarks on Cavalry* (1771), for example, Brent Nosworthy observes that "the methods of resolving combat on the European battlefield remained much the same between 1715 and 1740," but that in the following two decades "a new style of warfare would emerge, one which placed much greater importance on manoeuvre and grand tactical finesse." Whether the principles of this new style emerged from practice, or vice versa, "in 1750, 'tactics' was a newly emerging concept" on the modern battlefield and would become a governing concept in the evolution of modern military writing.[34] Many works on cavalry tactics followed, often connecting tactics, now more broadly defined, to equitation and dressage.

Captain Robert Hinde (1720/1–1786), Royal Regiment of Foresters (light dragoons) and a contemporary of Pembroke and Tyndale, for example, opened his comprehensive *Discipline of the Light-Horse* (1778) on the origin of British light-horse in 1745 and its transformation into light dragoons. Turning to dressage and equitation—to "instructions for managing the horses and teaching the light dragoon to ride" ("managing" here means *manège* training)—he not surprisingly commends Pembroke as a guide. Hinde then links the training of horse and rider directly to tactics. "But as light troops are more intended to act loose than in bodies," he writes, "their principal practice should be to acquire personal address, viz. to manage the horse well, to use the sword with dexterity, and fire the carbine with great justness. . . . Light troops so formed and exercised, may be used as dragoons were formerly." As mounted troops capable of dismounted action, "They are to be employed in reconnoitring the enemy, and discovering his motions . . . made use of for distant advanced posts, to prevent the army from being . . . surprised by the enemy, [and] sent out to distress the enemy"—precisely the duties that writers a century later would invoke as the raison d'être for cavalry as a modern arm.[35]

Continental writers took the same tack. In the posthumous *Mes Rêveries* (1757), or *Reveries, or Memoirs upon the Art of War* (also 1757), for example, Maurice de Saxe (1696–1750), marshal general of the French Army, took a panoramic view that included tactics—one based on a doctrine of mobility particularly applicable to a system of tactics for light cavalry. Like Enlightenment thinkers in all fields, Saxe sought to put the "art of war" on a scientific basis by identifying its fundamental principles and placing them in a rational order. Some years later, Emanuel von Warnery (1720–1776), general in the Prussian Army, published, in French, his

Remarques sur le militaire et le marine des Turcs et Russes (1771), or *Remarks on Cavalry* (1798 and 1805). Like Saxe, Warnery regarded mobility as the sine qua non for cavalry, but he turned the question of mobility to speed and then to "the greatest force of cavalry . . . *rapidity and vivacity*" (my italics), or "its most essential and important advantage." Consequential in the shock of heavy horse, rapidity was critical to the maneuvers of light horse. Theorists of both cavalry tactics and military equitation, Saxe and Warnery, like their contemporaries, put the horse controlled by the horse soldier, under all conditions, at the center of their thinking.[36]

In *Vorlesungen über Die Taktik der Reuterey* (1818), or *Lectures on the Tactics of Cavalry* (1827), finally, Warnery's successor, Friedrich Wilhelm von Bismarck (1783–1860), likewise based tactics fully on mobility, though his interest for us may lie elsewhere. If Hinde exemplified Enlightenment thinking, that is, Bismarck embodied the early nineteenth-century Romantic zeitgeist, nowhere more clearly than in his depiction of the genius of the ideal general. Most generals, Bismarck proposed, command either the "art" of tactics or the "science" of strategy, but "the *union* of art and science in one general, forms the perfect commander" (my italics) . . . *almost.* The perfect general *also* must possess a "natural gift . . . the talent to seize, as it were with one glance, the advantages and disadvantages which may arise from the situation of ground or troops, and to single them out from all other objects (*coup d'oeil*); this characterises the man born to become a general" and defines "*military genius!*"—as it also would for Lewis Edward Nolan.[37]

Eminent Victorians

In the introduction to a recent reissue of Nolan's *Cavalry: Its History and Tactics* (1853), Jon Coulston points out, "Apart from official manuals coming out of Horse Guards . . . there were very few cavalry treatises written in the first half of the nineteenth century by British authors."[38] They more than made up for it in the second half. Writers on cavalry in Victorian England, prolific and recondite, rivaled on the page the scale and scope of the British cavalry in the field, and they appear to have suffered even fewer setbacks. Cavalry performance in the Crimean War and Indian Mutiny provided military planners and writers at mid-century with opportunities for reviewing and reforming current theory and practice. Less happily, cavalry performance in the Second Boer War of 1899–1902 provided those at the turn of the century with lessons more negative than positive, particularly regarding horsemanship and horse management, resulting in scathing reviews and significant reforms.

An officer who writes books

In an age filled with dashing, colorful, and controversial cavalrymen, Captain Lewis Edward Nolan (1820?–1854) likely holds pride of place.[39] Following a commission in the Austrian Imperial Hussars, Nolan joined the British 15th Hussars and served with distinction in India, while also "making his name as a superlative horse-master, swordsman and steeple-chase rider." Next posted to Crimea, Nolan entered history in the "charge of the light brigade" at the Battle of Balaklava in October 1854. Joining the 17th Lancers for the charge and killed by fragments from a Russian shell, Nolan was the first British casualty, his death generally described in highly romantic, often hagiographic terms. Paraphrasing A. W. Kinglake's massive firsthand history, *The Invasion of the Crimea* (1863–1877), for example, Anglesey writes, "From [Nolan's] throat there came an unearthly shriek, and as his frightened horse swerved back through the 13th Light Dragoons, Nolan's sword arm, no less than his impeccably gripping knees, remained uncannily frozen in the positions they had held at the moment of death." Nolan also played a role in the miscommunication of orders that led to the British debacle, and the military, the press, and the public either eulogized him as gallant or pilloried him as goat, depending on interest and point of view.[40]

While Nolan's posthumous advocates and adversaries alike agreed on his gifts as horseman and swordsman, they differed on his personal attitude and behavior, regarding him either as a self-confident or as an arrogant officer, who, in the sneering assessment of Lord George Paget, "writes books, and was a great man in his own estimation," a sentiment quoted, but also softened, by Henry Manners Chichester in the *Dictionary of National Biography*. In point of fact, Chichester added, the accomplished Nolan "spoke five European languages and several Indian dialects." Nolan also excelled at music, and, despite his short life, wrote two books salient in the literature of mounted warfare: *The Training of Cavalry Remount Horses, A New System* (1852) and *Cavalry: Its History and Tactics* (1853)— the former a thorough but neither doctrinaire nor pedantic Baucherist treatise on military dressage and equitation, and the latter a widely cited history of cavalry and survey of tactics and a polemical, if diplomatic, statement of Nolan's opinions on the current and future states of cavalry.[41]

An early British advocate of Baucher, Nolan immediately credits the French master with the "invention" of the new system of dressage that Nolan lays out in *Training of Cavalry Remount Horses*. Nolan's *New System*, more specifically, comprises an adaptation and translation of Baucher's thinking and writing into an applied system for British cavalry informed by Nolan's experiences training cavalry remounts on Baucherist principles. "I had at first intended translating [Baucher's *Nouvelle Méthode*] from the French," Nolan writes, "but experience showed me that certain modifications were necessary to adapt it to the use of our cavalry."[42] Following preliminaries, the work unfolds a systematic program of lessons for the dressage of military horses and equitation for the military riders who must effect that dressage: two lessons, totaling ten days, "On the Snaffle"; plus seven lessons, totaling fifty-four additional days, for "Horses Bitted." The prefacing endorsements from fellow officers both emphasize the short time required by Nolan's system to train a horse and commend the uniformity of its results.

Though Nolan as pitchman promotes his "new system," in *Training of Cavalry Remount Horses*, on the brief training period that both saves resources *and* improves preparation, Nolan as horseman defends it on the soundness of the training itself, training that instills self-confidence and mutual confidence in horses and riders alike, and that also makes use of veteran horses to instill confidence in recently acquired remounts. Based "on a few simple principles," the system enables "the rider at last to reduce his horse to perfect obedience," while preserving the horse "from the effects of bad temper in the rider." As Nolan cautions, "Of all bad tempers in horses, that which is occasioned by harsh treatment and ignorant riders is the worst." He takes as axiomatic that the military horse must be supple and in hand, for "A horse not up to the bit, is unfit for cavalry duty . . . and in warfare totally useless in a mêlée or single combat." He thus advocates *manège* training: "My object . . . is merely to detail that work which requires the man to have patience and 'tact,' and which shews decided intelligence in the horse," and *manège* work is the surest means to that end.[43]

A year after *New System*, Nolan published *Cavalry: Its History and Tactics*, "an extraordinary statement on theory and practice of the mid-century period," in Jon Coulston's apt phrase, "the product of long and deep reflection by a consummate horseman who loved horses and profoundly respected the military service." Since few "special books exist . . . treating exclusively of cavalry, and none certainly of any importance in the English language," Nolan writes in his preface to *Cavalry*, he will offer "such suggestions as he [the author] hopes may assist in bringing forward this important arm to the level of the intelligence of the age." Disclaiming thoroughness or originality, he simply will organize and illuminate what others have said, together with his opinions. While his first four chapters indeed sketch a fairly prosaic history of cavalry and survey of its current tactics, the following ten chapters offer "all that I have to propose as a New System, or as a partial improvement upon the old one." System aside, Nolan's basic point is clear: the right men with the right arms, trained with an emphasis on "individual efficacy" and inspired to be "bold, resolute, and rapid," will defeat even superior numbers: "individual prowess, skill in single combat, good horsemanship, and sharp swords, render all cavalry formidable."[44]

Given Nolan's training with Baucher and his own expertise as "a riding-master for years," he naturally devotes a full chapter to military riding. Like his French contemporaries d'Aure and L'Hotte, Nolan wanted to meld *manège* and cross-country riding, though he came to the challenge from a British point of view. British officers, Nolan argues, continue to ape the once fluid but now "stiffened" Continental style of equitation when riding in military drills, though "none of our dragoon or hussar officers would for a moment think of riding across country [when riding to hounds] in a foreign seat, or in any other way in the manège fashion." He thus urges against "copying this seat and system from the foreign riding-schools" and for taking "example from our bold cross-country riders." Yet, he also exhorts, "Our cavalry now is wanting in its most essential qualification—'riding.' It is not sufficient that a dragoon can sit his horse; he should be completely master over him." Among other reasons, "charges resolve themselves into mêlées . . . and the unfortunate fellow who cannot manage his horse is lost." The cavalryman, in short, must cultivate the finesse of the true European school rider while showing the boldness of the British cross-country rider.[45]

Nolan would reappear twice in the public imagination a century later, as cinematic icon rather than as military author. Michael Curtiz's interwar romantic adventure, *The Charge of the Light Brigade* (1936), framed by Tennyson's romantic poem and set in a history more fanciful than factual, parallels public state conflict with private romantic triangle; and, in both plots, it embodies integrity and loyalty, as against perfidy and betrayal, in the person of Major Geoffrey Vickers, played with dash by Errol Flynn. In a key plot device, the film replaces the historical miscommunication of orders accidently conveyed at Balaklava with fictional battle orders intentionally forged by Vickers, in effect cinematically absolving British military command of ineptitude that led to six hundred casualties. Nostalgic for the British empire, although an American production, the film employs the flawed but ultimately heroic and vindicated Vickers as its vehicle for affirming imperial reach, might, and rectitude. While its secondary characters include Lords Lucan and Cardigan, by name and suitably whitewashed, its central character Vickers, a conspicuously good horseman as well as the agent of the misguided charge, clearly alludes to Lewis Edward Nolan.[46]

Tony Richardson's corrosive Vietnam Era military satire, *The Charge of the Light Brigade* (1968), by contrast, applies a veneer of historical detail and accuracy to its skewering of the British general staff in the Crimea. Richardson portrays the historical Lords Raglan, Lucan, and Cardigan, the latter of whom led the charge, as arrogant aristocrats whose family connections and landholdings enabled their rise through the military ranks despite their incompetence, and whose class snobbery, cowardice, and knavery conspired to justify their eventual decision to blame the fiasco on the dashing professional soldier Captain Nolan, dismissed by Raglan, in the film, with the self-damning assessment: "He rides too well. Knows a lot, but he has no heart." Absent as an identified figure in Curtiz's version, Nolan is the lynchpin in Richardson's telling: the film opens with a shot of Nolan mounted, and closes with a montage that crosscuts shots of Nolan's corpse, trampled in battlefield dust, with shots of an obviously symbolic wounded horse being put down by a trooper, while Raglan, Lucan, and Cardigan, offscreen but audible, blame one another for the disaster, not yet having settled on Nolan as their scapegoat.

British and German reformers

Captain Valentine Baker (1827–1887), commander of the 10th Hussars from 1860 to 1873, also served in Crimea, admired Nolan's *Cavalry*, and followed it with his own *The British Cavalry. With Remarks on its Practical*

Organization (1858), a short book that nonetheless canvasses the state of play across a range of recent developments in British cavalry from arms, organization, and maneuvers to horses, horsemanship, and horsemastership.[47] Baker opens his case, "Cavalry forms one of the most important branches of military service; for without this arm, victory loses half its results, and defeat is rendered doubly disastrous." In thus assigning a primary value of cavalry not to charging *in* battle, but to pursuing an enemy's retreat or protecting one's own retreat *following* battle, Baker shifts the arena of operations from battlefield to environs, a shift that would affect both horsemanship and swordsmanship and the teaching of them. More important, however, he seeks to understand why British dragoons had not proven more effective in the Crimean War. Presuming the superiority of British men and mounts to those of other nations, he concludes that "the system" of training and deploying horse soldiers and their mounts must be to blame. Reforming it is the subject and goal of his book.[48]

Baker's proposed reforms, as noted above, range across almost all aspects of cavalry. To invoke only those pertaining directly to horses and horsemanship, Baker praises the British thoroughbred and the "various breeds of great excellence" in the colonies, but also disparages the current state of British breeding and advocates state studs as the best "means of producing a class of animals suitable for military purposes."[49] Moreover, since breeding suitable animals and then rendering them unserviceable would be folly, he also proposes changes to equitation and tack that will preserve the horses once in service. He advocates Nolan's principles of military equitation, for example, together with a British "hunting saddle seat . . . for the dragoon" (as opposed to "the old German seat") and "a saddle shaped as a hunting saddle" and shorn of complications. As for bits, he likewise advocates simplicity of design that recognizes the practical constraints on training recruits as horsemen: "We cannot expect dragoons to have good hands as a body, and therefore we should so bit the horses, that they may be under perfect management in the hands in which we are compelled to place them."[50]

A decade later, Lieut.-Col. George T. Denison (1839–1925), Canadian national and nationalist, and more notable as reformer than as officer, published *Modern Cavalry: Its Organisation, Armament, and Employment in War* (1868). On the premise that recent improvements in weaponry necessitate changes in tactics, Denison writes "with the view of advocating certain alterations in the organisation, armament, and employment of cavalry in modern warfare." More specifically, as he elaborates, the universal *principle of progress* "applies more to the art of war than to any other subject," but "cavalry in most countries are now armed with almost the same weapons as when Alexander the Great used them against the Persians." Cavalry, instead, should take advantage of "improvements in fire-arms" and of "the employment of mounted men trained to fight on foot with these new weapons." Armed with carbines for dismounted action, and revolvers for mounted, cavalry will enter the modern age. Denison concludes, uncommonly and primarily on the basis of Confederate tactics, "There is no doubt that the revolver is beyond all odds the best pistol for mounted forces of any description. . . . [It] is apparently the most deadly arm that has ever yet been invented. . . . The sword, lance, carbine, long rifle, or cannon, do not have the same murderous effect."[51]

Denison subsequently appealed to history to bolster his position. In *A History of Cavalry from the Earliest Times, With Lessons for the Future* (1877), he argues that armies initially used horses for their speed only, "the warriors dismounting to fight on foot; [but] soon the idea arose of employing man and horse as a projectile weapon"; at the same time, though, armies recognized that the "mobility and speed" afforded by the horse also made cavalry the ideal arm "for scouting, reconnoitring, raiding, &c." While the cavalry charge against infantry using horse and man as projectile rendered cavalry formidable through the Napoleonic Wars, "the rifled firearm" effectively precluded such action in the future: "The bravest cavalry, charging in the boldest manner, are almost certain to be cut to pieces by the terribly destructive fire of the breechloading rifle." For that reason, and because of parallel improvements in cavalry's own firepower, the arm has chosen to exploit its unique fitness for the detached duties of security and information. "The chances are that in the next European war," Denison predicts, "whichever nation employs the mounted rifles extensively will be found winning decisive campaigns by the wise adoption of a necessary reform."[52]

Many years later, following the Boer War and on the eve of yet another European conflict, Denison published a second edition of *A History of Cavalry* (1913), using a preface to the second edition to point out that "thirty-six years ago the prejudice among the professional cavalry officers against the use of firearms by cavalry was universal and very extreme";

that he had "predicted" in *A History of Cavalry* in 1877 that cavalry comprising mounted rifles would be the successful cavalry "in the next great war"; and that this prediction, in fact, only repeated "what I had said in my book on 'Modern Cavalry,' published in 1868." The Franco-Prussian War, he had written in 1877, proved the "inefficiency of the sabre as an offensive weapon" compared with the deadly efficiency of the revolver, and "the Boer war of 1899–1902 and the Russia-Japanese war of 1904 have confirmed . . . the necessity for the adoption of modern firearms by the mounted soldier." With that, Denison leapt into the *arme blanche* versus fire action controversy then waging, and, with some justification, claimed vindication for his early advocacy of the latter.[53]

In *Notes on Cavalry Service* (1873), Captain John Cecil Russell (1839–1909) joined the reformers advocating small light cavalry formations armed with modern carbines.[54] Writing, like Denison in 1877, in the wake of the Franco-Prussian War of 1870, Russell rejects claims that cavalry cannot influence "battles of the future" because "the modern rifle and field-piece" preclude successful charges and attacks, advocating instead "making [cavalry] more fit to meet the new conditions of combat. Mobility in seeking for opportunities of action, and intelligence in taking advantage of them when found, are more than ever required." Specifically, the modern cavalry carbine raises the "question in what situations and to what extent fire can be profitably made use of, without interfering with the duties and qualifications of the mounted arm." He answers, one, that cavalry and infantry—horse soldiers and foot soldiers—differ in training and action, so a new corps of *mounted infantry* or mounted riflemen is needed for rapid transport of firepower; and two, that cavalry can make effective use of "their carbines without . . . losing the qualities proper to their own service," namely, the service of screening and reconnaissance. In those capacities, however, "The dismounted men must never be allowed to forget that they are cavalry soldiers, and that they have only quitted their horses for a special temporary purpose": mounted *or* dismounted, in a word, the modern cavalryman remains a *horse*man with "cavalry spirit."[55]

Colonel Frederick Chenevix-Trench (1837–1894) also joined the reformers in *Cavalry in Modern War* (1884), a wide-ranging technical manual written expressly for officers. A veteran of the Indian Mutiny and officer in the 20th Hussars, Chenevix-Trench wrote *Cavalry in Modern War*, the most influential of his four books on military topics, while serving as military attaché at St. Petersburg. Like Baker, Chenevix-Trench took "certain grave defects in the present organisation of our cavalry" as his point of departure; and, like Russell, he devoted considerable attention to the theory and practice of dismounted service. Surveying British efforts to develop cavalry's strategic and tactical capabilities, for example, he observes that the American Civil War "was destined [not only] to draw the attention of the military world to this subject anew [but also] to be looked upon as a turning point, or the beginning of a new chapter in the history of what cavalry may be trained to achieve." Chenevix-Trench had in mind, among other things, the use of carbine and revolver, the tactic of detached raids against the enemy's lines of communication, and, especially, the deployment of dragoons or mounted rifles.[56]

Though Chenevix-Trench neither subscribed to "the theory and belief that the day of cavalry on the actual battle field is virtually over," nor conceded the obsolescence of the charge or the saber, he argued that while the improved infantry rifle reduced cavalry's role on the battlefield, the improved cavalry carbine increased the arm's "range of duties and . . . sphere of independent action."[57] British cavalry now must take advantage of its ability at any time to fight on foot as well as on horseback, and, more specifically, it should study the American Civil War for lessons on successful dismounted tactics. Union and Confederate horse soldiers, in brief, owed their effectiveness not only to the breech-loading cavalry carbine and the revolver, but also to the tactics enabled by them. "The annals of the American Civil War teem with instances proving [one] that their horsemen were able to do what it is so often affirmed that European cavalry have failed in—viz. to act with boldness and skill both on foot and on horseback"; and two, "that cavalry may be used precisely in this manner and with the most signal and decisive results."[58]

Important and closely related works on cavalry by two German contemporaries of Chenevix-Trench and veterans of the Franco-Prussian War, Major-General Carl von Schmidt (1817–1875) and Prince Kraft zu Hohenlohe Ingelfingen (1827–1892), also appeared between 1875 and 1885: the posthumous *Instruktionen des Generalmajors Carl von Schmidt* (1876), comprising Schmidt's papers, published in English as *Instructions for the Training, Employment, and Leading of Cavalry* (1881); and Ingelfingen's *Ueber Kavallerie* (1885), published in English as *Letters on Cavalry* (1889).[59] Written during this critical period of transition from mounted to dismounted action as the primary role of cavalry, both works sought to transform constraints on cavalry that threatened its demise into opportunities for cavalry that would confirm its necessity and ensure its future. Both were defensive rhetorical maneuvers, and both arrived at the same position.

Schmidt's wide-ranging work opens with the assertion that "technical improvements and inventions" have far less importance than "a mental, intellectual dimension [described by] *handiness, mobility, manoeuvring power, rapidity, independence,* and *lightness*." The first of them, handiness, "the greatest dexterity of the individual mounted soldier in handling his arms, depends on a good, sound system of equitation, upon the training of man and horse, and a perfect understanding between them." Although Schmidt discusses military equitation at length, however, he also must reconcile his ideal of the horse soldier as horseman with the reality of the horse soldier as foot soldier. "The dragoon must not be or become a mounted infantry man," he writes, "but when he cannot attain on horseback the object indicated to him, he must be able to accomplish it on foot with his firearm." Put differently, the dragoon fighting in his primary role as mounted horseman uses "the horse, rapidity, and mobility" as his "chief weapons; after these comes the sabre." In his secondary role, however, the dragoon "must be able to fight effectively on foot under all circumstances" and "exchange the sabre for the carbine." Schmidt is arguing, like Russell, that the dragoon must be able to act as an infantryman, but not become one, if he is to retain his identity as a cavalryman and cavalry its identity as an independent arm.[60]

Although Ingelfingen, by contrast, considers an independent cavalry arm a "Utopian fantasy," he too seeks to transform current constraints on cavalry into future opportunities that argue for the arm's necessity. The horse is the key. The infantry's "improvement in firearms," he argues, extended the battlefield and, thus, the distance and speed required for approaches to charges "at the more rapid paces." That increase, plus an increase in the length of marches, plus the "demand that cavalry shall be able to carry on systematic dismounted action," weakens the physical state of cavalry horses and calls for more rigorous conditioning of them. The demand for dismounted action, moreover, requires breeding and training horses for that capacity: "At one time squadrons used to use only certain selected horses for patrols since most of them would refuse to leave the ranks. Now every horse has to be trained to go by itself in open ground." In short, changes in cavalry's role, whatever the gains, tell on cavalry's most critical asset: the horse. Though clearly the future, moreover, dismounted action will tell on the morale of horsemen and compromise the arm "if a trooper once believes that he can fight without his horse," or if the army converts dragoons into "mounted infantry. This new creation will, like all hybrids, be little use, and will fail entirely to discharge the most essential duties of a good cavalry."[61]

American Vistas

American cavalry writers, like their British counterparts, were absent in the first half of the nineteenth century and conspicuous in the second half. The United States had little history of mounted warfare prior to the nineteenth century, and only recent history of formal cavalry, the arm established tentatively in the late 1790s and permanently in 1832. European conflicts provided the material for mid-century American writers, while the Civil War provided postwar planners and writers with sufficient history and lessons to occupy them for the next several decades. These analysts included the controversial Union general George B. McClellan; the influential military educator Arthur L. Wagner; the Medal of Honor recipient and prolific author William Harding Carter; and the atypical lower-ranking officer Jonathan Boniface.

The first modern war

"There is little mention of military horsemanship during the American Revolution of 1776 or the War of 1812," Phil Livingston and Ed Roberts point out in *War Horse* (2003), for the simple reason that the former was fought primarily on foot and the latter primarily on the sea. "In 1792, Congress authorized the formation of a single company of dragoons, mounted infantry, a second in 1796 and a full regiment in 1798. . . . At the end of the war [of 1812] . . . Congress cut the mounted arm . . . and in 1816, abolished it completely." Picking up the story in *American Military Horsemanship* (2005), James A. Ottevaere adds, "The United States Army fielded no regular cavalry from 1815 until June of 1832, when Congress authorized a six company Battalion of Mounted Rangers."[62] Since the United States, as a result, lacked not only a body of theory on military dressage and equitation but also one on cavalry organization and tactics, the army turned to Europe, when the Civil War loomed, for models. Captain Kenner Garrard (1827–

1879) published *Nolan's System for Training Cavalry Horses* (1862), a redaction of Nolan's *Training of Cavalry Remount Horses* of 1852, and, at about the same time, Major-General George B. McClellan (1826–1885) published his report, *European Cavalry, Including Details of the Organization of the Cavalry Service among the Principal Nations of Europe* (1861).[63]

McClellan was posted as military commissioner to Europe in the 1850s, and, in that capacity, he observed armies fighting in the Crimean War, traveled to military establishments across the Continent, and reported his findings in the pre–Civil War *European Cavalry*.[64] The publisher's preface touts the value of such comparative studies "in so important a science as that of war, and in so momentous a period as the present. . . . Organizing for the first time a large and splendid cavalry force, to meet a want not felt in times past the United States Government is determined to introduce into the system every thing worthy of imitation in European cavalry." In addition, the book "forms a valuable companion-volume to the *Field Regulations for the United States cavalry in time of war*," an unsurprising assertion since McClellan also wrote those regulations. McClellan shows that European cavalries, despite national differences, shared fundamental principles. One of these, dating to the Renaissance, held that wielding a weapon and simultaneously controlling a horse were, in effect, one inseparable martial art. Another, also with long history, reflected the debates on horsemanship in Europe in the 1850s: though differing in emphasis, European cavalries concurred on providing instruction in *manège* riding for troopers "limited to the actual necessities of the service, and by the natural capacity of the horse."[65]

Toward the end of the century, Colonel Arthur L. Wagner (1853–1905), instructor at the U.S. Infantry and Cavalry School and a military reformer, opposed two articles of faith: one, that officers must learn by experience, with academic study at best a supplement; and two, that only officers with "genius—an innate talent for war—could comprehend and apply the principles of strategy and command." Wagner argued, instead, for the necessity of academic study and for the possibility of teaching strategy and command to capable officers even if not innately gifted with genius. He thus wrote two textbooks for the Infantry and Cavalry School. *The Service of Security and Information* (1893) concerns the use of "covering detachments" to protect (or secure) a main body, whether infantry or cavalry, from surprise attack by the enemy; and the use of detached patrols to reconnoiter (or gain information on the enemy's position and intentions). The much broader and heftier *Organization and Tactics* (1894) surveys how an army should organize cavalry, what tactics a modern cavalry should use, and whether the organization employed enables or constrains execution of those tactics.[66]

The two volumes align. In the former, Wagner promoted extensive use of detached units and concluded that "the two elements of security and information," screening and reconnaissance, are "inseparable" and so can and should be performed by the same detachments—a view rejected by Bernhardi, Douglas Haig, and Alonzo Gray. In the latter, he argued that improvements in infantry firepower having reduced opportunities for battlefield charges, developing a capacity for "dismounted fire action gives to cavalry an independence and a power which add immeasurably to its value"; a lack of that capacity, conversely, would result in cavalry being "reduced to a condition of dependence upon the infantry, and relegated to the rôle of a purely auxiliary arm." Cavalry's future, consequently, lay in dragoons, "troops armed [with sabers and carbines] and trained with a view to fighting effectively either mounted or on foot." The dragoon, or "the American cavalryman of 1864–5 is the type to which all European mounted troops are . . . approaching."[67] Needless to say, Wagner considered the important roles of horses and horsemanship, either explicitly or implicitly, in this future, but his focus lay in cavalry mission, organization, and tactics. Prewar reformers like McClellan had looked to Europe for adoptable models, but postwar reformers like Wagner promoted American innovations as models for Europe.[68]

Contending forces

While officers like McClellan and Wagner wrote directly for army staff or school use, "some officers, such as Captain William H. Carter and 1st Lieutenant Jonathan J. Bonifice," Ottevaere points out, "wrote manuals, books and instruction guides for publication outside the Army in hopes that their ideas and methods would find a hold within the service." A decorated veteran of four wars and an effective reformer in military organization, Carter (1851–1925) published a number of military works in multiple genres, including *Horses, Saddles and Bridles* (1895). Less celebrated, Boniface (1875–?) served in the 4th U.S. Cavalry and appears to have written one book, *The Cavalry Horse and His Pack* (1903). Unlike the many works from the period on cavalry organization, strategy, tactics, or weaponry, or even on military dressage and equitation, these two works share a focus on cavalry *horses* and on a myriad of practical matters pertaining to equine acquisition, equipment, care, and use.[69]

Noting that *Horses, Saddles, and Bridles* "is not intended as a treatise on equitation," and believing that a warhorse did not need advanced *manège* education, Carter nonetheless stressed "the value of the riding school as a means of bringing all the men and horses to an average state of efficiency."[70] More concerned with horses than horse soldiers, in any event, and particularly with cavalry saddle horses, he contends, "Remounts for cavalry must have certain qualifications, the most important of which are the possession of sufficient mobility to execute tactical maneuvers at varying degrees of speed and the ability to stand hard service while carrying great weight." Military horses, of course, were not only saddle horses, and Carter also sets forth the dissimilar but equally important qualifications in artillery horses, particularly those serving in the horse rather than field artillery. "The artillery horse is a combination of the saddle and the draft horse," he writes, and a team of horse artillery horses, in particular, must be able to pull a gun at cavalry speed while ridden by a gunner. Providing both traction and transportation, they "cannot be simply saddle animals, but must possess good draft qualities."[71]

Boniface intended his weighty "compilation," *The Cavalry Horse and His Pack*, to provide "the young cavalry officer one volume embracing [all] the duties and responsibilities which confront him." Taking in fact a more focused view, however, Boniface produced probably the definitive tract on military equestrian travel, and, like Carter, he paid less attention to the soldier than to the horse, "without [whom] military history would be barren of those stirring events with which we are all so familiar, and which make us look forward to the achievements of our cavalry in all future wars." Since the American army makes no distinction between light and heavy cavalry, and "the regulation cavalry horse is the same for all regiments," Boniface reasons, the cavalry horse should combine the "spirit, agility, and intelligence" of a thoroughbred, with the "large bone and strong sinew, short back, stout forehand, strength, and hardihood" of a draft breed, but it should entail neither the weight of the latter nor the expense of the former. While "the American Cavalry service," he notes, "is light cavalry in all its work and training, in the weight of its men, and its armament and pack, and therefore requires light, strong, agile horses," cavalry is already "the most expensive branch of an army," and mounting it with thoroughbreds "would be too enormously expensive an experiment to try."[72]

Though Carter and Boniface focus on cavalry horses, neither altogether ignores the cavalry tactics and other actions enabled by them. "The development of the modern rifle," Carter observes, "led theorists to proclaim that frontal attacks . . . were things of the past and that cavalry must henceforth be relegated to reconnaissance and orderly duty"; and, on the whole, he does not disagree. And Boniface, identifying all American cavalry as light cavalry, concurs: "In mounted shock-action heavy cavalry finds its proper field, while in all the arduous varying duties of patrolling, screening, reconnoitering, raiding, escorting, and scouting, light cavalry stands pre-eminent." Carter and Boniface were confirming, in effect, what was largely a fait accompli by the turn of the century: the shift from battlefield maneuvers in massed regiments to reconnaissance and security duties in detached patrols. McClellan had pointed to the signs of this shift in his discussion of "skirmishers"; Wagner devoted a book, *Security and Information*, to it; and their followers would encode the new prevailing role as doctrine and, in the Great War, extend it to the "cavalry of the clouds."[73]

Great Expectations

The turning of a century provokes in most disciplines many works either retrospective, prospective, or both. So far as cavalry is concerned, the turn from the nineteenth to the twentieth century was no exception. British military historians such as Sir Evelyn Wood and Frederic Maude took long views backward, as did George Denison with his revised *History of Cavalry*. Soon after, the cavalrymen Alonzo Gray, Sir Douglas Haig, Captain Loir, and M. F. Rimington took nearer views, gleaning lessons from the American Civil War, Franco-Prussian War, and, especially, Boer War, to prepare young officers, and themselves, to wage the next European conflict looming increasingly close on the horizon. The Boer War served as a particular goad to the British, both in its catastrophic wastage of horses and in its clear lesson that the long ongoing shift from *arme blanche* and shock action to firepower and the detached duties of security and, especially, information, had taken a decisive turn. Put differently, as the American Malcolm Wheeler-Nicholson observed, "The British learned in the Boer War what we learned in the Civil War." That turn informed the vigorous "*arme blanche* controversy" featuring Bernhardi and Childers, and Generals John French and Earl Roberts.[74]

Long views

Nineteenth-century writers on cavalry advised officers striving for excellence to study treatises, old and new, on the theory and practice of their arm. Over the course of the century, they added *history* to theory and practice, as evidenced by works such as Sir Edward S. Creasy's *Fifteen Decisive Battles of the World* (1851) or Denison's *History of Cavalry*, as well as the many books, such as Nolan's *Cavalry*, that included one or more historical chapters. As Badsey writes in *Doctrine and Reform in the British Cavalry*, "Part of the military thought of the age was that a study of military history was believed to yield 'lessons' directly relevant to serving officers. . . . Part of the reasoning behind this belief was that the horse, sword and lance had not changed in any fundamental way for millennia, and neither had many of the basic duties of cavalry."[75] Since England, unlike the United States, had a long history of mounted warfare in general and light-horse in particular, British cavalry "believed" far more than American in the efficacy of historical writing. Two British military writers stand out: General Sir Evelyn Wood and Colonel Frederic Natusch Maude.

General (later Field Marshal) Wood not only enjoyed a distinguished military career but also wrote several books, notably *Achievements of Cavalry* (1897). Wood notes that he had published six studies in the *United Service Magazine* in 1892 "for the assistance of such of my young comrades as are not fond of close reading in military history," and subsequently had added six new studies bringing the original series up to the Battle of Rezonville in the Franco-Prussian War. Comprising the twelve studies, *Achievements of Cavalry* offers both military history and battlefield case study. It also concludes with an advocacy piece, "Mounted Infantry."[76] On the theory that England, "in wars of the future," must use horses "to convey infantry soldiers to [where they must] fight on foot," Wood reiterates his plea, made in 1874, for establishment of "a corps of Mounted Infantry . . . attached to a cavalry division in the field." When acting alone, it will form the "moving Base" of a cavalry division ahead of an army corps; when acting in concert with other arms, it will serve most usefully as "an escort to Horse and Field artillery." In either case, in "Savage" as opposed to "European warfare," conducted "over extensive tracts of country," mounted infantry "will prove of immense advantage."[77]

A more prolific author though less distinguished soldier than Wood, Frederic Maude (1854–1933) wrote *Cavalry Versus Infantry* (1896) as a study of tactics in the Napoleonic era, though it addresses other subjects as well.[78] Maude advances his idée fixe, for example, that British cavalry training undermines courage and discipline by overvaluing life: "The Germans believe that the first object of military instruction is to teach the soldier to know *how to die*, and we teach him *how to avoid dying*." He also advocates a military horsemanship comprising mastery of both *manège* equitation and cross-country riding. Citing as a contemporary exemplar General Heinrich von Rosenberg, "one of the finest all-round horsemen Germany has ever possessed . . . a born cavalry leader and keen soldier . . . one of the best steeplechase riders Germany has ever produced," Maude, more centrally, invokes two pamphlets written in 1815 and 1816 by the Prussian general Friedrich August Ludwig von der Marwitz, who had proposed that applying the moral and physical strengths required in battle "is infinitely more complicated [in cavalry than in infantry], because they depend on the cöoperation of two living organisms, the rider and the horse . . . the ultimate unit." This entails an art of horsemanship, "now extinct in our cavalry" and leading to the arm's decay, a proposition that Maude finds timely.[79]

In the subsequent and historically broader *Cavalry: Its Past and Future* (1903), Maude advanced a polemic for horsemanship as being essential both to future cavalry and to the future *of* cavalry, as well as against past and, more urgently, present efforts to exchange saber and shock for fire power.[80] As Maude argues, one, soldiers whose skilled horsemanship "conditions 'cohesion' [in shock action] gives also endurance and mobility" to the horse; and two, soldiers "taught to rely on the firearm, not on the sword [begin] to look on the horse as a mere means of locomotion, and not, as it really is, an essential part of the ultimate cavalry unit." The Boer War demonstrated "the supreme importance of mobility" to that unit; and mobility, to be efficient and effective, presupposes mastery of horsemanship, or "unconditional obedience" of horse to rider. Pivoting from horsemanship, he comes to his purpose, namely, to address "the pressing question of the day. 'In the future is it to be cavalry or mounted infantry?'"

Since Maude believes that hybrid troops cannot be properly trained, "All our efforts now should be directed to securing under *all conditions* the maximum possible degree of mobility, and from this point mounted infantry, or rifles . . . promise no advantage over orthodox cavalry."[81]

Nearer views

Many books published early in the twentieth century took up that question, and related questions, with reference to recent rather than historical wars. Journalistic "chronicles," for example, provided stirring firsthand narratives of the Boer War, such as Major Charles Sydney Goldman's beefy *With General French and the Cavalry in South Africa* (1902). Technical studies on tactics often used maneuvers or employed real or hypothetical problems as illustration, such as Major-General Haig's *Cavalry Studies* (1907); while others drew on a single recent "great war," such as Captain Loir's *Cavalry* (1912, 1916), illustrated by the Franco-Prussian War. Overviews of the cavalry arm, such as Major-General Rimington's *Our Cavalry* (1912), carried forward a genre reaching back to the eighteenth century. The technical studies and overviews typically shared two traits, identifying, one, young officers as their target readership, and two, instruction in "practical" or "applied" knowledge as their objective.

With General French and the Cavalry in South Africa typifies chauvinistic Boer War narratives.[82] Inferring from South Africa, for example, that "the conditions of modern warfare are likely in the future to make even greater demands on the cavalry, and . . . to increase their sphere of usefulness," Goldman (1868–1958), war correspondent for the *Standard*, implies (and promises to document in his narrative) that the British cavalry, in particular, was up to the task, adapting its methods "to entirely novel and abnormal conditions of warfare [while] its mettle, its spirit, and its irresistibility remained unchanged and unimpaired." In "The Future of Cavalry in War," one of multiple appendices, Goldman also takes issue with critics who contend that "not the charging lancer but the well-posted rifleman will decide the battle of the future," reducing the role of cavalry to that of mounted infantry. Defining cavalry as "formed bodies of men armed and trained to fight on horseback and on foot," Goldman finds the arm uniquely suited to the critical duties of security and information. Rather than mounted infantry "equipped solely to fight on foot," he calls for "the man who is equally at home in the combat on horseback or on foot. In fact to deprive your mounted troops of the possibility of action mounted is simply to deny them an advantage with no compensating gain."[83]

A few years later, between major commands in the Boer War and, more famously, the Great War, Sir Douglas Haig (1861–1928) published *Cavalry Studies* (1907).[84] Based on "Five Staff Rides which took place under my direction when Inspector-General of Cavalry in India (1903–6)," *Cavalry Studies* constitutes "a carefully compiled study of certain imaginary Cavalry operations," as opposed to one based on actual military events.[85] While academic study produces "a knowledge of Military History," participation in "War Games" and "Staff Rides" develops "our power of decision."[86] Given the unique capacity of cavalry for the principal "services" of information and security, and the critical "functions" of reconnaissance, pursuit, and threatening or cutting off the enemy's line of retreat, "the rôle of Cavalry, far from having diminished, has increased in importance." Its leaders, therefore, must be at once sufficiently knowledgeable to understand what they see and sufficiently decisive to act forcefully on that understanding. Moreover, Haig argues, modern warfare entails "*large Armies [that] entail large numbers of Cavalry*," and its leaders, therefore, must be trained in mass tactics.[87] Since *Cavalry Studies* focuses entirely on tactics, including mass tactics, it simply assumes the value of horses and horsemanship and mentions them at most in passing.

While Haig's *Cavalry Studies* posed imaginary problems to be solved on field maneuvers, two other works posed problems to be solved at a desk: Major J. H. V. Crowe's *Problems in Manoeuvre Tactics, with Solutions for Officers of All Arms, After the German of Major [Julius] Hoppenstedt* (1905); and *Exercices pratiques de cadres* (1911), by Colonel Auguste Pierre Monsenergue, translated as *Cavalry Tactical Schemes: A Series of Practical Exercises for Cavalry* (1914). The former, as Crowe quotes Hoppenstedt, was designed "to assist officers in self-instruction in Manoeuvre Tactics . . . and to familiarise officers with the action of the three Arms when working in co-operation." The latter, as General Gough writes in an introduction, was designed to challenge the officer being tested with "concrete tactical problems [that require him under time constraints] to arrive at a decision—*to do something*." Specifically, Crowe poses brief problems and provides lengthy solutions, while Monsenergue reports problems posed, responses made, and critiques of the responses.[88] Though differing in pedagogical tactics, in short, Haig, Crowe, and Monsenergue all share mission and strategy.

Across the Channel from Haig, Captain Loir published *Cavalerie* (1912), translated as *Cavalry. Technical Operations. Cavalry in an Army. Cavalry in Battle* (1916), a dense, technical book outfitted with twelve very large and detailed folding maps, based on the Franco-Prussian War.[89] Like other writers, Loir identifies reconnaissance as "the essential function of cavalry" at present, and he negotiates between shock and fire action without precluding either of them by proclaiming the former the primary raison d'être for cavalry and the latter a secondary role that it plays well, a conventional position by 1912. Noting that the maneuvers regulation states, "Cavalry will engage on foot . . . when the tactical situation or the nature of the ground prevents it from fighting on horseback," he concludes that there should be no "occasion for argument, and the question, 'Sabre or carbine,' ought not to present itself"—the problem lies not in weapons or tactics, but in extremist theories regarding them. Loir also pays much attention to the elemental cavalry asset—the horse—not only as a vehicle for marches, "the basis of all operations of war," but also as a weapon in battle. To sustain "moral force" when fighting, then, the trooper must have confidence in his horse and realize "that between his knees he can always feel a strong, handy, game and gallant comrade."[90]

The Prussian general Gerhard von Pelet-Narbonne (1840–1909) had published an earlier book also on the single (and signal) example of the Franco-Prussian War: *Der Kavalleriedienst im Kriege* (1901), or, as translated, *Cavalry on Service, Illustrated by the Advance of the German Cavalry across the Mosel in 1870* (1906). Intending "to discuss the action of cavalry on service in an applied form based on the actual events of a war," Pelet-Narbonne had written not "a history," but a work of "practical instruction." In *Cavalry in the Russo-Japanese War: Lessons and Critical Conversations* (1907), soon after and by contrast, the Austrian count Gustav Wrangel also drew on a single war, but less for practice than theory. Posing the question of mounted saber versus dismounted carbine, as would Loir, Wrangel answered, "The ideal would perhaps be for [cavalry] to do each equally willingly [and efficiently]," but then, judging this impossible, he also took the conventional position: "As long as we lay principal stress on good dashing horsemanship and the clever handling of the *arme blanche*, and relegate training with the rifle to the second place, so long shall we foster the offensive spirit of our cavalry. On this stands and falls the whole activity of the cavalry arm."[91]

Major-General M. F. Rimington's *Our Cavalry* (1912), finally, exemplifies works aimed to provide young officers with an overview of cavalry operations and tactics.[92] An unreconstructed soldier of the "older school," Rimington advocates mounted shock action over dismounted fire action and, thus, steel over firearms, and hews to the "*doctrine . . .* of the resolute offensive," whether aimed at infantry or cavalry, that shock action entails. Rimington also devotes considerable attention to horses and horsemanship, launching the theme, with some fanfare, in his preface: "Those who have never felt the sensation of a really good horse bounding and stretching away under them, and the consequent elation, the wonder as to 'what could stop us?' cannot grasp what a cavalry soldier's feelings are in the 'Charge.'" He extends the theme, moreover, to horsemastership. Officers, that is, not only must teach newer officers how "to ride and to train a horse," and how to teach their men, in turn, to do so, but also must instill in troops, directly and indirectly, the critical importance of horse care: "From the day [the recruit] joins, no opportunity should be lost of teaching [him] that amongst his first duties is to love, honour, and have a pride in his horse." The reason is simple: "the care of the horse is the weak link in the cavalry chain" that caused massive equine death in the Boer War and the current need for reform.[93]

Arme blanche controversy

"The experiences of the Boer War," Spencer Jones writes in his study *From Boer War to World War* (2013), "set the stage for a heated debate . . . throughout the pre–First World War period as to what tactical role cavalry would play in any future conflict." It pitted "those who favored dismounted firepower as the principal tactic, championed by Lord Roberts, [against] those who preferred shock action, headed by John French," though, as Jones notes, both schools "emphasized flexibility and . . . the creation of a hybrid soldier." Generals French and Roberts also served as seconds, as it were, to the publishing duelists Friedrich von Bernhardi, zealous German reformer and theorist, and Erskine Childers, Anglo-Irish author with service in South Africa. The duel was fought, from one side, in Bernhardi's *Cavalry in Future Wars* (1899, 1906, 1909) and *Cavalry in War and Peace* (1910, 1914) and, from the other side, in Childers's *War and the Arme Blanche* (1910) and *German Influence on British Cavalry* (1911). Childers's books were rejoinders, specifically and respectively, to Bernhardi's books.[94]

"The Art of War," Bernhardi opens *Cavalry in Future Wars*, "has been revolutionized since the Franco-German War." Though he argues, specifically, that cavalry has gained importance through the "wider sphere of activity," including "reconnaissance and screening," enabled by dismounted fire action, and though he approves such action, he nonetheless stands with *arme blanche*: "Having admitted that dismounted action has increased considerably in importance . . . it nevertheless remains the fact that the combat with cold steel remains the chief *raison d'être* of the Cavalry." In *Cavalry in Future Wars*, wider in purview than the tactical *Cavalry in War and Peace*, Bernhardi also emphasizes horsemanship and the "improved methods of breaking and of equitation [that] must bear fruit in every branch of their activity [including, above all] the conservation and endurance of the horses themselves." Alluding to mid-nineteenth-century debates in cavalry circles, and clearly placing himself in the German tradition, Bernhardi advocates training that secures "the most perfect harmony between man and horse." Rather than *haute école* airs, he calls for the practical though still refined movements that military horsemen such as Nolan had advocated.[95]

With no apparent interest in military horsemanship, Childers focuses only on weapons and tactics. In *Arme Blanche*, he proposes that "a pure type of mounted rifleman [be] substituted for the existing hybrid type," because the latter in fact, and inconsistently, rests on "a theory of tactics derived from the steel."[96] Its proponents, namely Bernhardi and French, hold dear a "purely sentimental conservatism" that treats *arme blanche* as "a religion in itself," or, as Childers puts it in *German Influence*, "a matter of faith, not of reason; of dogma, not of argument; of sentiment, not of technical practice." Though recognizing mobility as the cavalry's chief asset, they fail to see that "the primary distinction between the horse-soldier and the foot-soldier lies in the horse, not in the weapon carried by the man"; they fail to grasp that "the sole issue is, by the agency of what weapon can the horse, in conjunction with the will and the manual skill and strength of the man, be used to the best advantage?"—and they thereby fail to ken the obvious answer: the rifle. For them, "Cavalry spirit" is the "subjective idea" and the "terror of cold steel" its "objective corollary," or, put differently, "Cavalry spirit, in its inmost essence, means the spirit of fighting *on horseback* with a steel weapon, in contradistinction to the spirit of fighting on foot with a firearm." In a word, Childers concludes, they confuse "mobility and combat," and, as a result, what counts as "cavalry spirit."[97]

Across the pond

Captain (later Colonel), Alonzo Gray (1861–1943), 14th U.S. Cavalry, published *Cavalry Tactics as Illustrated by the War of the Rebellion* (1910) on the premise "that there is no modern principle of cavalry tactics, which is accepted today as correct by any first-class military power, which was not fully illustrated during the War of the Rebellion."[98] Employing countless extracts, "mostly from the Records of the Rebellion," together with commentary, Gray draws out principles and lessons illustrated by the war. Not surprisingly, those principles and lessons pertain to the current controversies in the 1910s regarding mounted duty with saber versus dismounted duty with carbine. In Gray's view, however indispensable the carbine is to dragoons when fighting dismounted, it holds little, if any, value in mounted action, where the saber and revolver are the weapons of choice. Moreover, though Gray retrieves any number of graphic firsthand accounts of soldiers "sabering" their adversaries—"cutting them down" or "running them through"—he concludes, based on many other accounts, "In the individual combat the revolver will be the winner in almost every case." Gray's lessons regarding the revolver, based on its extensive use by the Confederate cavalry, skew the simple binary formula of saber versus carbine, though only in theory, that drove the British controversy.[99]

Five years after Gray, and two years before the United States joined the Great War, Paul Trapier Hayne (1846–1921), captain, 12th Cavalry, presented a series of three lectures on cavalry at the Army Service Schools, focused respectively on reconnaissance, other duties, and combat, and published as *Lectures on Cavalry* (1915). Though Hayne discusses neither horses nor horsemanship, his tactical preferences for reconnaissance, "the most important use of cavalry," and for raiding, the important innovation of the Civil War that "in future wars . . . will get great results," presuppose a particular horse and rider unit—a light, agile, and fast mount under a capable cross-country rider. Hayne also serves to illustrate how fully the concept of hybrid cavalry—dragoons fighting mounted with saber and dismounted with carbine—had become doctrine, though his precise formulation— "dismount to fight on foot only when it appears that a mounted charge would not succeed or would not accomplish the mission of the cavalry"— also illustrates the continued prevalence of a conservative and institutional bias toward charging with steel as embodying "cavalry spirit."[100]

Two years after Hayne, and one month before the United States entered the war, Major Leroy Eltinge (1872–1931), also speaking at the Army Service Schools, presented one of three lectures, "Notes on Cavalry," subsequently published in *Notes on Infantry, Cavalry and Field Artillery* (1917).[101] "In the fortress warfare that is going on to-day along the whole line in France," Eltinge writes, "there is no use for cavalry, but the hope of driving the enemy from his trenches and thus have a chance to pursue him to his destruction leads both sides to hold cavalry close in rear of their intrenched lines." Describing other "uses," he cautions against one of them, the British practice on the Western Front of using cavalry as infantry. In Eltinge's view, "To take cavalry from the important duties which it alone can perform and use it as infantry is to throw away the advantage of having cavalry." Those duties, "except when in emergencies [cavalry] is used as infantry," all require mounted action. Subscribing to the same current orthodoxy as Hayne, Eltinge advises subalterns, "As a general rule for your guidance you may take this: Stay on your horses and utilize your mobility as long as you can. Dismount to fight on foot when you have to do so or can gain an advantage thereby."[102]

After Cavalry—What?

The Great War arrived, and theoretical controversies led to real consequences. The Great Powers slogged it out in the mud of the Western Front, where the role of cavalry, after some important early mounted actions, settled largely into that of dismounted infantry; but cavalry also rode it out, as it were, in the vast open spaces of the Eastern Front, where horse cavalries, in putatively their last great engagements, clashed in mounted charges against one another as well as against infantry, and, when appropriate, took dismounted action. As Colonel Oliver Spaulding of the U.S. artillery wrote of the Great War in general, "It is difficult to avoid over-estimating the changes due to this war." Given the unprecedented size of guns, volume of fire, and immense scale, "At first sight it almost seems as if a new art had come into being. . . . This war is like no other in history in that it is being fought not by armies but by nations."[103] Soon after, and with much controversy, the United States joined the conflict.

Confirming what Bernhardi and Haig had anticipated, this conflict of mass, mechanized armies would close some options for cavalry action, but also open others. Hewing to the doctrine of mobility as the foundation of cavalry, and to the idea that providing information and security was now the primary role of the arm, Bernhardi advocated at length, during and especially after the war, a combined effort of mounted cavalry and the air service, the cavalry of the clouds, as the future of reconnaissance. Concurrently, the highly visible military writer B. H. Liddell Hart, predicting the tank as the primary cavalry mount of the future, maintained mobility as axiomatic, but eliminated the horse as its provider. The interwar years witnessed the increasing and inevitably total mechanization of cavalry, and this, in turn, provoked elegies, typically in the form of memoirs, for the passing of the horse and, with him, the horse soldier.

The man you love to hate

While European countries decimated one another from 1914 to 1916, the United States maintained a state of neutrality until April 1917. Fearing well before then that the country likely would not escape entanglement in the European, or "foreign," conflict, American military thinkers explored broad policy questions that directly or indirectly affected cavalry doctrine and strategy, training and tactics, and human and equine resources. Three books published in 1915—Frederic Louis Huidekoper's *The Military Unpreparedness of the United States*, William Harding Carter's *The American Army*, and Maxwell Van Sandt Woodhull's *West Point in Our Next War*—all began with the premise that the United States, wary of militarism and protected by two oceans, was unprepared for war with a foreign adversary. Like England, and unlike Germany, the United States had neither a standing army at wartime establishment nor a state horse breeding program, and so, on declaration of war, would lack the trained men and horses needed to field a cavalry. As these thinkers warned, the United States would learn in 1917 what England learned in 1914: neither cavalrymen nor cavalry horses can be mobilized, trained, and deployed quickly in large numbers.[104]

Needless to say, American military thinkers were not alone at that moment in worrying questions of national policy. As Carter had done, General von Bernhardi extended earlier views on cavalry strategy and tactics to include their broader context. In two prewar works published in English translation in 1914, *Germany and the Next War* and *How Germany Makes War*, Bernhardi amply confirms the characterization of him by A. Hilliard Atteridge, the editor of the English abridgement of *Cavalry in*

Future Wars, as "best known in England as a writer of the 'Jingo' School which has done so much to produce the war." Indeed, the two newer books gave Bernhardi, ironically, a kind of negative celebrity in the world of anglophone military intellectuals as "the man you love to hate," to coin a phrase.[105] Those books also set the stage for Bernhardi's postwar *apologia pro vita sua*, *The War of the Future: In the Light of the Lessons of the World War* (1921).

To describe *Germany and the Next War* as "jingoist" is an understatement. An egregious exercise in nationalistic belligerence, the book begins with the premise, "The value of war . . . has been criticized by large sections of the modern civilized world in a way which threatens to weaken the defensive powers of States by undermining the warlike spirit of the people."[106] In response to that criticism, Bernhardi aims "to prove that war is not merely a necessary element in the life of nations, but an indispensable factor of culture, in which a true civilized nation finds the highest expression of strength and vitality." While advancing this agenda, Bernhardi isolates a point where national policy meets military strategy, and he pivots from it to strategy and to the tactics that support it. Reprising his earlier affirmations of "the great value of [cavalry's] operative mobility . . . notwithstanding all modern weapons," he now argues, "The reconnaissance and screening duties of the cavalry must be completed by the air-fleet. Here we are dealing with something which does not yet exist, but we can foresee clearly the great part which this branch of military science will play in future wars."[107]

In *How Germany Makes War*, Bernhardi reinforces the jingoism of *Germany and the Next War*, both in warning Germany about the British–French alliance threatening it and in admonishing Germany "to renounce all weakly visions of peace, and eye the dangers surrounding us with resolute and unflinching courage." With respect to strategy and tactics, he continues to derive precedents from both the American Civil War and the Boer War; he reaffirms that "cavalry is now itself equipped with firearms, and can use them when charging is impossible"; he identifies reconnaissance and screening as among the most important duties of current cavalry; and he concludes "that in almost all its spheres of action the importance of cavalry in war has very much increased with the growth of armies, though its employment differs somewhat from that of former times." Bernhardi also steps up his advocacy of new technologies. He acknowledges that "it is scarcely possible to gauge to-day the probable effect of aerial navigation on the future conduct of war;

for it is almost always impossible to discern the full significance of new inventions and innovations"; but he also promotes "the latest achievements in technics, namely, *aeronautics*," for use in reconnaissance and screening.[108]

Now that the wars were over

For many decades, each major advance in firepower not only heralded "the end of cavalry" but also, ironically, led to the new tactics and weaponry that ensured cavalry's vitality. The introduction in the late nineteenth century of "the new generation of magazine rifles, practical machine-guns, and quick-firing artillery using new explosives and smokeless propellants," for example, fed military angst, but it also led to a sometimes overlooked countermeasure: the cavalry's embrace of the machine gun. Prewar writers who advocated equipping cavalry with automatic weapons professed that, as a result, horse soldiers would gain firepower while machine guns, reciprocally, would gain mobility.[109] The theory proved correct, though the myth that the destructive power of the machine gun in the Great War marked the end of cavalry took hold in the public imagination and still has purchase, despite revisionist debunking.[110] The machine gun, and the debates around it, can serve as metaphor for mechanization in the Great War in general.

General von Bernhardi, for example, did not fade away either with age or with the armistice, though he modified his prewar tone and outlook in the postwar study, *The War of the Future* (1921). While his focus here is more ideological than technical, his technical conclusions illustrate a shared postwar outlook on the implications of new technologies and associated tactics for the cavalry, specifically the necessity for dismounted action and the importance of aircraft for reconnaissance. Bernhardi claims, one, that the war not only confirmed what he had forecast, namely, "the duties of reconnaissance have been almost entirely taken over by aircraft," but that it also signaled something new: an air force, like a land cavalry, can conduct reconnaissance "effectively only when the supremacy of the air has been fought for and won"—or, put differently, only when the cavalry of the clouds engages and vanquishes the opposing air cavalry. Bernhardi also claims, two, that cavalry in the war of the future, even more than in the past, will fight on foot, and will use "the speed of the horse . . . no longer for attack but for rapid strategic maneuvers. . . . A tactical dismounted unit must be the foundation of the whole organization." For increased efficacy on foot, "mounted troops should have machine-gun squadrons."[111]

Known to history primarily as a writer of pulp fiction and pioneer of comic books, Major Malcolm Wheeler-Nicholson (1890–1965), American author of *Modern Cavalry* (1922), believed with Bernhardi that air service would play a role in reconnaissance, but not that aircraft or "trucks and lorries" would replace the horse for other duties.[112] As devoted to the horse as Rimington, Wheeler-Nicholson treats horsemanship and horsemastership, at length, as no less important to the modern cavalry of fire action and aerial coordination than to the traditional cavalry of steel and shock. An advocate of "mounted sports," particularly foxhunting and polo, as ideal for honing the skills and proving the mettle of officers, he proposes, "No officer who is not an enthusiastic horseman has any place in the cavalry"; and, likewise, no trooper who is a poor horsemaster has any place in it. The officer must have a "love of the horse and all that pertains to him," and the soldier "must be made into an enthusiastic horseman who develops enough affection for his mount to be willing and anxious to take every care of it." Finally, Wheeler-Nicholson urges a change in expression from "training the horse" to "training the horse for war," with a commitment in practice to eliminating anything superfluous to that end, including, for instance, advanced dressage.[113]

In counterpoint to Wheeler-Nicholson, Captain B. H. Liddell Hart (1895–1970), the apostle of mechanized cavalry, published a seminal essay in the *Atlantic Monthly*, "After Cavalry—What?" (1925). Mobility defines the "essential value" of cavalry, Liddell Hart argues, so rather than think of cavalry "as men on horseback," one should think of it "as the mobile arm." This eliminates "many misconceptions and prejudices," since cavalry "in fulfilling its historical functions . . . has assumed many different forms and comprised radically different types and patterns." More important, this will cause "general staffs of all nations" to cease regarding the tank as an "adjunct to the infantry." The tank, rather, is the cavalry's new mount, supplanting the horse, a tank force is *"the modern form of heavy cavalry,"* and tank warfare is that cavalry's new mission. The correct tactical use of tanks follows from those premises: "to be concentrated and used in as large masses as possible for decisive manoeuvre against the flanks and communications of the enemy."[114]

That heady essay became a lynchpin of Liddell Hart's book, *The Remaking of Modern Arms* (1928), whose "keynote . . . is MOBILITY—of movement, action, organization, and, not least, of thought." Liddell Hart had claimed in "After Cavalry" that the tank would result not only in "the rescue of mobility from the toils of trench-warfare, but with it the revival of generalship and of the art of war." Elaborating that claim in *The Remaking of Modern Armies*, he argues that a modern form of mobility can cure "the paralysis of the last war"—the immobility of trench warfare—and prevent it from recurring in a future war. That form comprises a new technology, "the cross-country six-wheeler" equipped with a machine gun, together with the new tactics that it enables—a form presaging "the coming of tank armies" and, with it, the end of mass foot infantry and the beginning of "an era of offensive mobility."[115] In the end, Liddell Hart skewed his vision when he reimagined horse and rider, too literally, as a "one-man tank . . . built of ordinary car components," deployed in groups that "can be controlled like a squadron of cavalry or company of infantry." Unlike many other writers, however, Liddell Hart had vision, whether perfectly clear or not.[116]

Remains of the day

Scholars of the Great War traditionally tend to agree that mounted cavalry played a major and decisive role on the Eastern Front, but a minor and indecisive one on the Western Front, a consensus reflecting laudatory early accounts of British action in the East and culminating in the supposition that the Palestine campaign of 1917 was "the glorious swansong of the British horse-soldier," the specific conceit of the "swansong" invoked, for example, by Brereton and Coulston.[117] This supposition, in turn, entailed a complementary proposition: "The First World War was the last gasp of the British cavalry arm," and "the last world conflict in which cavalry horses played a major role," in the words, respectively, of Anthony Dawson and of Phil Livingston and Ed Roberts. In fact, British and American horse cavalries, however etiolated, soldiered on for years, though they moved steadily from horse to mechanization in the 1930s and decisively in the 1940s, as we shall see in the next chapter. The cavalry *arm*, of course, is another matter.

While the anglophone powers shrank their horse cavalries in the 1930s and fielded few mounted troops in the Second World War, Germany in the 1930s "was building up its horse strength for the cavalry, artillery and transport of the Wehrmacht," and, together with Russia, represented "the two most important cavalry forces that survived with a horse-mounted combat capability into [the war]," fielding millions of equines between them. Notwithstanding those numbers, the era of horse cavalry was ending, as "large formations of war horses and riders," in DiMarco's nice quip, "took the field one last time in a conflict that also included jet aircraft and nuclear weapons."[118] The epoch of the cavalry arm, however, and its doctrine of mobility, continued for decades, its horses largely if not wholly displaced first by tanks in the Second World War and then by helicopters in Korea and Vietnam.[119] None of this, however, suggests that nations consigned their horse cavalries to the hell of the forgotten. On the contrary, a plethora of Great War and mid-century memoirs suggest that nostalgia for horse cavalry ruled the day.

General Lucian K. Truscott (1895–1965), for example, published his Second World War memoir, *Command Mission: A Personal Story* (1954), eleven years before his death in 1965. During that period, Truscott also composed a second memoir left in manuscript, edited by his son, Colonel Lucian K. Truscott III, and published as *The Twilight of the U. S. Cavalry: Life in the Old Army, 1917–1942* (1989). In an author's preface, Truscott called his account neither "the biography of an individual," nor "the history of the cavalry during its last years," but, rather, "a faithful portrayal in outline of cavalry *life* as one cavalry officer and his family lived and loved it in the stations and on the posts in which he served"; or, in the words of Edward M. Coffman, Truscott gives "the reader a well-written, meticulously detailed portrait of cavalry life as it neared extinction." As its Wagnerian title suggests, the portrait has tragic accents, but is fundamentally elegiac in substance and tone—an elegy for interwar horse soldiers and their horses, not unlike Paul Rodzianko's *Tattered Banners* (1939) and Vladimir Littauer's *Russian Hussar* (1965).[120]

"Riding and the care of animals," Truscott writes, "constituted the principal items of instruction in the mounted-service schools" in his time, including "the use of cavalry weapons mounted—that is, the use of pistol and saber. . . . Mounted work was by far the most popular part of the [cavalry] course. . . . It was the horse . . . that distinguished the cavalryman from the other branches. . . . It made the cavalry an elite branch of the service, and it was in horsemanship that most of the legends of the [Cavalry School] originated." Truscott himself confers the cachet of legend on the "'gay young cavalrymen' willing to try anything, and never willing to back down on any dare," including a group of drunken, tuxedo-clad young officers who took leave of a formal dinner party to "ride bareback, with only halters to guide the horses over five-foot jumps." He also confers it on the history of cavalry in the interwar years: he recites the mechanization of cavalry as if it were a series of battles, with a foregone outcome, fought by a succession of legendary senior officers. These included three devoted horsemen who, nonetheless, acknowledged mechanization as the future, and one chief of cavalry who "was bitterly opposed to mechanization," who "believed in *horse* cavalry," but who, as a result, misread the political context and, ironically, helped bring about what he most feared.[121]

Lieut.-Col. E. G. French's *Good-Bye to Boot and Saddle, or The Tragic Passing of British Cavalry* (1951), by contrast, aspires to tragedy rather than elegy. An avuncular defense of the ancien régime and a dyspeptic assault on modernity, the work celebrates horse cavalry and rails against its mechanization, while it canonizes French's father, General John French, and denounces his defamers—both that mechanization and defamation, closely linked in French's mind, being tragic in their consequences. A key to French's mind lies in his relentless use of three closely related terms and the values represented by them: *smartness, esprit de corps,* and *cavalry spirit.* Though manifest in uniforms and deportment, in brief, smartness extended far beyond them: not only a value in itself, smartness engendered esprit de corps and also "was essential to the maintenance of a high standard of discipline"; and both esprit de corps and discipline, in turn, were requisite to the "true cavalry spirit."[122] General Gough defines that spirit in a foreword as "a spirit of boldness, of independence, and a capacity for rapid decision"; French exalts it as an omnipresent and, in essence, incarnate force.[123]

In French's telling, cavalry between the Boer and Great Wars, possessing "cavalry spirit," comprised "good horsemen and horsemasters, skilled in the use of firearms as well as their traditional steel weapons." In the *arme blanche* controversy, however, their leader General John French faced "ruthless revolutionaries, who demanded the complete abolition of the *arme blanche*, and the transformation of cavalry into mounted infantry." French, his political allies, and their "troops" won the battle, but they lost the war. "And so," his son begins his final chapter, "we come to the sad story of the unhappy fate of British cavalry. . . . Our famous cavalry regiments, one after another, lost their beloved horses and found themselves converted into mechanical units." For French, the embodiment of mid-century nostalgia, "A cavalryman is a horse-soldier armed with sword or lance, augmented by some kind of firearm. If you withdraw his horse and his steel weapons he ceases to be a cavalryman, and nothing, not even pretentious nomenclature, can make him one." Armored cavalry might live on, but horse cavalry, over a century of regimental tradition behind it, had died.[124]

Only two hundred years separate the ascent of light-horse under Frederick the Great, with cavalry "at the zenith of its power," from the mounted arm's eventual demise following the Great War. Those two centuries, however, witnessed an extraordinary trajectory. "The system of Frederick the Great has served its purpose, and it is left to our arm now to look about and devise a new system to suit the present state of affairs," George Denison wrote in 1877, at the height of the evolution of cavalry from mounted shock action to its "new system" of dismounted fire action. Though contemporaries such as Prince Hohenlohe Ingelfingen, writing in 1885, perceived that evolution as quantitative—more sophisticated tactics, more intelligent men, more agile horses, more powerful weapons—we can see it, from our historical vantage, as qualitative. The transformation of horse cavalry into armored or air cavalry, the legacy of the Great War, may have carried cavalry mission and strategy, even "cavalry spirit," forward, but it left the *horse*, the definitive component of horse cavalry, behind. The military horse, cavalry and otherwise, however, already had endured a history at once compelling and dispiriting between the Boer and Great Wars.[125]

4

Remounts and Wastage

This was in 1914. A year later many of the men and horses I had
seen, and even some of the horses I had myself ridden, lay dead on
the battle-fields of France.

Lida Fleitmann Bloodgood
Hoofs in the Distance (1953)

The role of the horse, or remount, in modern warfare raises different questions than does that of cavalry: the latter questions concern the organization and operations of a mounted arm, while the former concern the
acquisition and deployment of the equine resources needed for that arm.
Cavalry and other mounted troops in the Great War of 1914–1918, and
the saddle horses used by them, as we have seen, commanded a good deal
of contemporary military and civilian attention, both official and unofficial. So did heavy and light draft horses and draft and pack mules. That
should surprise no one: cavalry at the time—before its subsequent mechanized and air incarnations—still depended on saddle horses, and field
and horse artillery units, as well as the army in general, depended on draft
breeds in even greater numbers.

The Great War, on its Western though not Eastern Front, was fundamentally an infantry and artillery war, not a cavalry war, though revisionist historians recently have stressed the critical role played by cavalry in
delaying German infantry during the early months of the conflict in 1914,
as well as other contributions made by the cavalry in the West through
1918.[1] The Great War, however, even on its Western Front, remained
largely a *horse* war. Military horses were bought, bred, shipped, trained,
deployed, wounded and nursed, and killed, buried, or cast, in numbers
that beggar the imagination. Millions served ably and bravely. Given the

material value of horses as both energy resources and weapons, government plans, commissions, and reports studied them as fungible assets; and given the symbolic value of horses as warriors, war workers, and veterans, countless posters, photographs, paintings, memorials, and memoirs treated them as intelligent and sensitive comrades-in-arms.

The Horse Question

While horses in warfare initially conjure images of officers and troopers charging en masse at speed, warhorses, particularly after the "golden age" of vast horse cavalries, primarily were *workhorses*—draft and pack animals—that provided transport, traction, and mobility for all three military arms—infantry, artillery, and cavalry. Horses provided not only the mobility that served cavalry action, but also the fuel that powered the entire war machine: draft horses transported supplies and munitions to battle; hauled guns and their accoutrements in battle; and carried the wounded and dead from battle. Saddle horses, draft horses, and mules enabled war, including the Great War, and arguably the Allied victory in it. As Field Marshal Sir Douglas Haig wrote in 1919, "If in March 1918 the equine force of Germany had been on the same scale and as efficient as the British equine force, the Germans would unquestionably have succeeded in breaking through between the French and British armies, and inflicted a defeat so great that recovery might not have been possible."[2]

Horsing an army

Needless to say, modern armies had faced challenges to the supply and ongoing resupply of equines in large numbers long before the Great War. The ruthless calculus of supply and demand in warfare applied no less to horses than to men: an army depended upon a national capacity not only for rapid initial mobilization of large numbers of quality saddle and draft horses, but also for ongoing replacements for the countless horses either killed, wounded, or otherwise rendered unable to serve further. This capacity presupposed a robust "horse supply," defined by Lieutenant Jonathan Boniface in *The Cavalry Horse and His Pack* (1903) as "the horse population of the country" potentially available to the military, plus a steady supply of "remounts," a remount being a "horse purchased to fill up the vacancy caused in a cavalry service by the loss of one of its horses." As Boniface goes on to explain, "The ability to raise and maintain a cavalry force, to a large extent," depends on the horse supply. "In time of peace the procuring of proper remounts is not a very difficult matter. . . . In time of war the subject of remounts becomes one of the most important as well as one of the most difficult to handle."[3]

The policies, politics, and practices for initial supply and ongoing resupply of horses differed across the Continent, Great Britain, and the United States—primarily on the basis of breeding programs, with "some countries," as Boniface observes, "supporting government breeding-farms for this purpose, others supporting or partly supporting studs only, while others procure their remounts by purchase." In the European tour that issued in the report, *European Cavalry* (1861), General George McClellan found that Russia, with the world's then largest horse supply and "the best mounted [cavalry] in Europe—certainly the best on the Continent— [maintained] no *haras* (breeding-studs) for the general service of the army"; and that France, except at Saumur, did likewise, though, "At each remount depot there are stallions of the race most suitable for crossing with the mares of the vicinity." Germany maintained a national stud system with "a central commission at Berlin charged with the regulation of the purchase of horses" through three national "districts" and "sub-commissions" and "several remount depots." Debates in the United States and Great Britain on breeding military horses took place against that backdrop.[4]

Boniface, for example, began his manual with the forecast that "the time for [the horse's] displacement is not yet, although evil prophets may proclaim the horseless age as already ushered in." Forty years earlier than Boniface's forecast, in the American Civil War, as Phil Livingston and Ed Roberts tally, the Union and Confederate armies together had fielded some 650,000 horses; and only five years earlier than Boniface, in the Spanish-American War, the U.S. military had dispatched some 45,000 horses and mules to Cuba and the Philippine Islands.[5] Though a census taken in 1903 counted 18,280,007 horses on farms and the range in the United States, only a small percentage were available for military service.[6] Boniface did not advocate a national breeding program, but he invoked with obvious admiration Germany's national stud system in order "to show the great pains taken in Europe to produce superior cavalry horses, and it may be taken as the best system in existence"—high praise not fully shared by General von Bernhardi. In the coming years, Boniface's countrymen would advocate for a national breeding system, though not for "government breeding farms."[7]

George McCullough Rommel (1876–1945), chief of the Animal Husbandry Division of the Department of Agriculture and one of those advocates, began a government report circulated in 1911, *The Army Remount Problem*, on the note then being sounded by all militaries: "The mounting of troops in an army is a most serious problem" in peace as well as in war, particularly since "horses multiply slowly, and a reserve must be provided." Acknowledging that the U.S. military satisfactorily maintains "the supply [of horses] necessary for . . . its present peace footing," he argued that it would need a reliable supply "in case of war" to grow from twenty thousand to fifty thousand cavalry horses "before a shot was fired or a saber drawn." Market forces, however, had produced an abundance of draft horses suitable for artillery but a scarcity of saddle horses suitable for cavalry—the British army in South Africa alone had purchased over a hundred thousand horses of "the right type" from the United States. Rommel intended, of course, not just to point out a problem, but also to describe at length—and to promote—Congress's steps in addressing it: creation of the Remount Service in 1908 together with approval of a plan in 1911 for a breeding program of government-owned stallions and civilian mares.[8]

Spencer Borden (1848–1921), a civilian horseman and author of *The Arab Horse* (1906), took a different tack in *What Horse for the Cavalry?* (1912). "Were we to become involved with any other power at the present time," Borden begins the latter, "the army of the United States knows not where to find horses for its mounted service, no matter what it should be willing to pay." While the national herd in 1910 numbered 24,016,024 horses, "these great numbers of horses do not represent animals available for the army." Moreover, as Rommel too noted, rearing and training take time, so "We cannot 'go shopping' for horses because we have plenty of money. Their supply must be the result of forethought and care." The United States, to its detriment, had not developed a national breeding program, Borden argued, cautioning, "Great Britain alone [among European nations] has been negligent in this matter, and has suffered the penalty." Two questions, in his view, thus presented themselves: one, should the government establish a breeding program, and two, if so, "what type of horses shall be bred?" Borden answers, one, indirectly, *yes, on the Hungarian model*, and two, implicitly but clearly, *the Arab horse*.[9]

Herbert H. Reese (188?–1960), who also served in the Animal Husbandry Division of the Department of Agriculture, published the bulletin,

"Breeding Horses for the United States Army," in 1918. Reese began with the problem, common to the United States and Britain, that "the right types [of horses] for cavalry and light artillery" had become scarce because farmers, responding to domestic market forces, turned from breeding light horses to breeding draft horses. In response, as Rommel too noted, Congress approved a program in 1911 for breeding government stallions to private mares, and, as Reese can now report, "the first appropriation for the remount breeding work was made available for the breeding season of 1913, making the first crop of 3-year-old colts available for inspection and purchase by the War Department in 1917." While he acknowledges, "The object . . . is to select for and breed sound horses . . . conforming to the cavalry or artillery types," he defends the program, as would other agriculturalists, and despite the just concluded war, primarily for its civilian value: "Notwithstanding the necessity . . . from a military standpoint, this work is nevertheless largely an agricultural proposition. . . . Good horse power is indispensable to successful farming, and good horses cannot be produced without good sires."[10]

The British history of centuries of global mounted warfare and cavalry action greatly exceeded an American history of horse cavalry begun in the nineteenth century and confined to North America. Debates about horsing the British army, accordingly, also had longer history and greater moment, particularly with respect to breeding. Opponents of "the establishment of a formal remount system" in England, Anthony Dawson points out, had argued since the eighteenth century that it "would have given the government a monopoly on the horse trade and therefore have been against the principles of Free Trade." Dismissing such objections, Lewis Edward Nolan countered in *Cavalry: Its History and Tactics* (1853), "It is building a house on the sand to organise cavalry without good horses. Government alone could work the necessary reform by importing stallions and mares of eastern blood, for the purpose of breeding troop horses and chargers for the cavalry of England." Valentine Baker concurred in *The British Cavalry* (1858), "No attention is paid by the Government to . . . producing a class of animals suitable for military purposes, and the supply is entirely dependent on private enterprise. Consequently, any sudden demand for cavalry remounts produces an immediate increase in price [or] deterioration in the quality." He, too, advocated government studs.[11]

Horsing the British army, an increasingly pressing problem in the ensuing decades, became critical in the Boer War of 1899–1902. Fought far from home, almost entirely by horseback, the Boer War not only required a steady supply of horses in great numbers, but also entailed shipping them by sea to South Africa from English and other distant ports.[12] Charles Sydney Goldman, on assignment *With General French and the Cavalry in South Africa* (1902), commented on the lack of mobility that hampered the British cavalry, "In nothing was adequate preparation more necessary than in the provision of a sufficiency of horses and a complete equipment for mounted troops." Elaborating in an appendix, "Remounts in War," Goldman made the familiar point, "One of the most important and most difficult problems the war has raised is the whole question of horsing an army in the field." The horse registration schemes introduced in the 1880s, for example, produced more horses suitable for draft than for saddle, and its purchasing officers, moreover, lacked appropriate experience in judging them. "The equipment of our cavalry [including horses]," he concludes, "raises one of the most important military questions of the day. There is no doubt that it needs entire remodeling."[13]

In the years immediately following the South African horse debacle, Sir Walter Gilbey (1831–1914), president of the Shire Horse Society and the Royal Agricultural Society and a prolific author of books on horse breeds and breeding, twice addressed the "Horse Question."[14] In *Horse-Breeding in England and India and Army Horses Abroad* (1901, reissued 1906), Gilbey argued that increased production of "the heavy draught-horses" needed for agriculture coincided with (indeed led to) decreased production of both light draft horses appropriate for artillery teams and saddle horses for troop mounts needed for a "Peace Establishment" of thirty thousand. Gilbey rejected the idea that "the British Government should embark upon costly horse-breeding operations in emulation of foreign powers," however, on the familiar if debatable grounds that "private enterprise in England" has produced superior animals: the problem is quantity, not quality. In this work, Gilbey limits himself to the proposal that breeding military horses to solve that problem belongs to "a civil department [of agriculture]; the business of procuring Remounts for troops, on the other hand, is essentially a soldier's task. The error lay in the attempt to combine the two"—a point similar to that made by Reese.[15]

Gilbey's *Horses for the Army: A Suggestion* (1902, reissued 1913), had more exigency: the demonstrable lack of preparation for the South African War, Gilbey writes in 1913, shows "that the system of obtaining remounts [in the Boer War] could not be depended on by the nation for future emergencies [and that] the situation has [worsened since]; the progress made by motor traction [has caused] tens of thousands of horses to be discarded . . . of types most essential for military purposes"—particularly types essential for horse and light artillery. Gilbey commends the Board of Agriculture's plan for "subsidising selected stallions for public service," but points out that it will not solve the problem of depletion in the national peacetime "supply of horses" available for military use due to exportation of horses to foreign markets. Moreover, the government's various registration and subsidy schemes may address the problem of wartime mobilization, but they cannot solve the problem of "loss and wastage in war." Instead of state breeding programs on the Continental model, he "suggests" complementing the existing registration and subsidy schemes with establishment of ten "Permanent Remount Depôts" for breeding government supported stallions with local mares.[16]

In *Cavalry Studies* (1907), General Douglas Haig had cautioned, "It must be borne in mind that the days of small Armies are past, and it is a simple fact that *large Armies entail large numbers of Cavalry*"—large numbers, in turn, that entail commensurately large numbers of horses. Acquiring them on the scale demanded by the Great War would prove more vexing than in the Boer War—a problem that post–Great War advocates of horse cavalry predicted would worsen in future wars.[17] Lieut.-Colonel R. M. P. Preston observed in *The Desert Mounted Corps* (1921), for example, "The disappearance from our English roads first of the coaches and then of the horse-drawn buses, has deprived us almost entirely of our once fine type of light draught horse, and it seems as if we shall, in the future, have to depend more and more on the Dominions for our supply of such horses."[18] Similarly, Major Malcolm Wheeler-Nicholson, the fervent American postwar defender of horse cavalry, called in *Modern Cavalry* (1922) for "the enlargement and use of cavalry recruit depots and . . . the augmentation of our remount service so that it can provide us replacements of trained horses in war time."[19] Liddell Hart, as we saw, did not concur.

As previously discussed, advances in firepower and automotive technologies during and following the Great War had rendered horse cavalry and, to a lesser but still considerable extent, overall military horse usage, increasingly anachronistic in England and the United States. "The mechanisation of the British Army during the 1930s," Janet Macdonald writes, "reduced the military requirement for horses, though mules were still used as pack animals in rough terrain." Likewise, James A. Ottevaere writes, when the United States Army, in 1936, integrated one field artillery and two cavalry units "into a new mechanized 7th Cavalry Brigade . . . scout cars, half tracks and M3 Stuart Light Tanks would replace their horses." As Livingston and Roberts add, "By 1944, the cavalry of the United States Army had been dehorsed and was being fought as mechanized infantry." The army's last horse mounted unit, the 287th M. P. Horse Platoon, was not disbanded until 1958, and the British Army's last pack mule unit, the 414 Pack Transport Troop, not until 1975. This, however, was all well after the fact of Great War equine carnage.[20]

A Tragic Waste

The "Horse Question" that vexed British and American military planners since the turn of the century comprised three problems: first, sustaining a military herd adequate for a peacetime establishment; second, mobilizing the large numbers of horses required for an initial wartime establishment; and third, ensuring the sustainability of those numbers given the wastage of horses in wartime. Though the specifics of those problems and their envisioned solutions differed between Great Britain and the United States, the two nations continued to share an obstacle not faced by Continental powers: neither maintained a state supported and administered horse breeding operation. Military planners, as a result, had to achieve two pairs of goals: one, compensate for reductions in desired equine types due to mechanization in the civilian sector, *and* incent civilian breeding of those types by creating markets for them; and two, register horses in the domestic civilian sector and ensure their availability for military requisition, *and* develop markets in foreign countries for the reliable importation of remounts in large numbers.

Like supply and resupply, wastage of horses—their loss by death, sickness, casting, or other causes—was a perennial problem with a long history. As Carter wrote in *Horses, Saddles and Bridles* (1895), "The loss of animals in all wars is very great, and occasionally the average is much increased by occurrences of an unusual nature." Wastage was high in the American Civil War and the British Crimean War and scandalously high in the Boer War, and, in all three cases, critics blamed poor horse management more than battle—the former being avoidable, the latter inevitable.[21] Nowhere was more contemporary and near contemporary blame laid on poor horsemanship and horse management than in the Boer War. Following a detailed indictment, for example, Goldman attributed not only wastage but also the army's poor performance, in effect, not to how horses failed men, but to how men failed horses; and both Rimington in 1912 and Wheeler-Nicholson in 1922 invoked (to quote the former), "the unfortunate ignorance in regard to horses, horsemanship, and horsemastership which . . . caused our bill for horses in South Africa to total twenty-two millions—that is, about one-tenth of the whole cost of the war."[22]

Long-distance marching, a major source of equine fatigue and lameness, had proven a problem in earlier conflicts. F. Chenevix-Trench had observed in 1884 that "*the* cause which does more than anything else to ruin the strength and efficiency of cavalry . . . is, it need hardly be said, sore backs"; and, since the combined weight of man and equipment was a primary source of exhaustion and sore backs, George T. Denison had advised officers even earlier to find ways to reduce any "unnecessary weight." With specific reference to the demands that the vast landscape of South Africa had placed on British cavalry, F. N. Maude wrote in 1903 that "the science of the preparation of horses for long-distance marching, and the possibilities of reducing the weight carried both in equipment and food for the rider, are entirely new discoveries." Having learned those lessons, Rimington and others would admonish officers to order men to sit erect and balanced when marching or briefly halted, to dismount at scheduled halts or other delays, and to dismount and lead their horses when feasible. Due to the equally vast landscape of the American West, U.S. cavalry had made those discoveries years earlier in the Indian Wars, lessons that also would serve them well in the "Punitive Expedition" against Pancho Villa in 1916.[23]

If deficient horse management was the main cause for equine deaths in South Africa, mass warfare was its multiplier. "Essentially a 'horse war,'" Graham Winton reminds us (quoting Thomas Pakenham) in *Theirs Not to Reason Why* (2013), the Boer War "swallowed horses as a modern army swallows petrol." The Army Veterinary Service and the Remount Service, both established in the 1880s, had been "totally unprepared and unsuited for the responsibilities expected of them [in 1899] in their first major conflict." As a result, they failed "to prevent the tragic waste of animals" that became a national scandal: Great Britain sacrificed some

325,000 horses and 51,000 mules out of a total of over 520,000 fielded.[24] Writers then and now have advanced many reasons for this high mortality ratio, the most salient, as noted above, being the *combination* of poor horse management and mass horse deployment.[25] As Winton observes, "Although there was some improvement towards the end of the war, horse-mastership was poor in all branches of the Army, including the regular cavalry and mounted infantry; the exception was the Artillery."[26]

Faced in the early teens with a potential conflict of unknown scale, scope, or duration, British military planners resolved not to repeat the equine debacle of the Boer War, but instead to devise viable plans both for mobilization of an adequate initial wartime horse establishment and for ongoing replacement of waste horses with fit remounts. Discussing German preparation in 1914, Bernhardi had been woefully mistaken in his declaration, "Motors . . . are the latest achievement in military transport service. . . . All draught animals can now be spared, which is of greatest importance." More prescient, the British anticipated correctly that motorized transportation and traction would play a major role in the coming conflict, but they also understood that draft and saddle horses in great numbers would be essential. As Kenyon observes, "The horse remained by far the dominant form of motive power on the Western Front (aside from railways)." The War Office strengthened both the Remount Service and Army Veterinary Service, and it beefed up its registration and subsidy schemes for private horses and motor vehicles. The Horse Census of 1912–1913, however, found some 589,401 horses suitable (though not necessarily available) for military service—an insufficient number that also included more heavy draft horses, and fewer light draft horses and horses of the hunter type, than were needed.[27]

Planners, put differently, were preparing for mobilization and ongoing resupply of equines in unprecedented numbers, to potentially multiple and far-flung theaters, for an unknown period of time. On declaration of war, the British peacetime establishment of 30,000 horses would have to balloon to an initial wartime establishment of 160,000. Though using a variant figure, Dawson adds the salient point that England "had to scramble to find '150,000 horses of the right sort' upon mobilization," while France and Germany, with national studs and breeding farms, mobilized 655,000 and 550,000 horses, respectively.[28] Lacking state breeding farms, as discussed above, the British after initial mobilization had to depend on commercial and domestic and foreign markets for remounts. Over the course of the war, the Remount Service purchased approximately 1,250,000 horses, 468,323 in the United Kingdom and roughly 780,000 from abroad (703,705 of the latter from North America).[29] Tens of thousands of light draft horses were in British service at any time on the Western Front, as were tens of thousands of saddle horses in the Mesopotamian and Palestinian campaigns.[30]

A total of 300,000 to 400,000 invaluable and highly praised mules joined them. Citing Major R. C. H. Berry's figures in the journal of the British Mule Society, Lorraine Travis reports that the British had 230,973 mules in service on August 31, 1918. During the war, in a chapter of *The Horse and the War* (1918) devoted to "The Gallant Mule," Captain Sidney Galtry wrote, "One of the many wonders of the war has certainly to do with the tens of thousands of mules transported from the Western to the Eastern Hemisphere and now actively pursuing the big part assigned to them in the Great Adventure." Enjoying what might be called a cult following, the mule was also the subject of much sardonic wartime doggerel, such as "Musings of a Mule," reproduced by Galtry, and "Verses to a Mule," included in a regimental publication, *Gas Attack of the New York Division* (March 2, 1918). Invoking a number of wartime encomia as testimony, Travis concludes, "A mule sees and hears more than a horse, is ever on the alert, is difficult to stampede, and makes a good sentry."[31] They were indispensable.

From purchase to active service, horses were highly vulnerable to disease, accident, and violent death. Winton calculates that 13,724 horses were lost on the transatlantic crossing alone, some 6,667 of them "through enemy action," mainly victims of German U-boats.[32] Once in service, horses faced the destructive power, as discussed previously, of quick-firing field guns, automatic machine guns, and magazine-fed rifles accurate at very long range.[33] Horse lines were easily visible by air, and, as Lucinda Moore observes, "The sheer surface area of larger creatures such as mules and horses meant that they were more vulnerable to flying shrapnel."[34] On the Western Front, artillery horses and troop mounts died primarily from shelling, but also from gas, drowning in liquid mud, or euthanasia. On the Eastern Front, "fearsome carnage amongst the horses" resulted from cavalry charges into artillery and machine gun fire. While mules were judged both before and during the war as less susceptible than horses to disease and fatigue, Goldman had been less sanguine in 1902: "The mule luckily is almost a freak of nature in its capacity for hard work on short rations [but] in the end they suffered, and heavy mortality overtook them."[35]

The scale of equine loss during the Great War boggles the mind and darkens the spirit. Estimated casualties totaled a staggering 8,000,000 for all belligerents, including 3,000,000 of the 6,000,000 fielded in total by France, Germany, and Austria. Wartime casualties of British equines alone reached 529,564 animals—losses that would have been far worse had the army not established seventy veterinary hospitals, curing some 395,000 of 552,000 horses treated as of February 1918.[36] Nonetheless, the Disposal of Animals Branch had to establish Horse Carcass Economiser units to facilitate the sale of horse meat and other parts from animals neither cured nor deemed serviceable, and, on a larger scale, also had to dispose of thousands of equine corpses, principally by burial in mass graves, though, when circumstances necessitated, by even more grisly alternatives, such as shooting mules inextricably stuck in mud, "and then as many as three men would stand on the corpses to sink them beneath the surface of the mud in lieu of a burial." As Ernest Hemingway would write in a related context: "All stories, if continued far enough, end in death and he is no true-story teller who would keep that from you."[37] For thousands of horses, that end was still to come.

On November 11, 1918, the British Army held 896,000 equines, a number greatly exceeding the needed postwar establishment. As a result, nearly 800,000 animals had to be disposed of, "as quickly and humanely as possible."[38] Many horses serving in France were returned to England for roles in agriculture or transport, but domestic mechanization was reducing demand for workhorses, so thousands still in France were euthanized, slaughtered for meat, or sold to French farmers. Horses who had served in Allenby's Desert Mounted Corps were even less fortunate. By government policy, no horses would be shipped from the Middle East back to England, Australia, or New Zealand. As a result, for example, "it was decreed that . . . some 20,000 [horses] belonging to the Yeomanry and others should be cast and sold off in Egypt . . . a sentence of protracted death in slavery." As Lieut.-Colonel E. G. French would write bitterly in *Good-Bye to Boot and Saddle* (1951), "Their services no longer required by an ungrateful country, these equine soldiers . . . were disbanded and dispersed no one knew where, many of them, it must be feared, to a life of wretchedness and overwork."[39]

The armistice that promised peace to humans had promised and delivered far less to their horses. The war would not stop reaching out from its grave, particularly for the thousands of horses suffering privation and servitude in the Middle East. At least one small light, however, glimmered through the gloom. A dozen years after the armistice, in 1930, the formidable and indefatigable horsewoman Dorothy Brooke arrived in the Middle East with her husband, Geoffrey Brooke, "posted as a Brigadier commanding the Cavalry Brigade in Egypt," a distinguished horseman and author of *Horse-Sense and Horsemanship of To-Day* (1924) and *The Way of a Man with a Horse* (1929). Dorothy Brooke raised some 40,000 pounds sterling for a War Horse Fund to buy and care for surviving and by then aged British military castoffs. By 1934, Brooke not only had rescued "some 5,000 ex-army horses," but also had founded the Old War Horse Memorial Hospital in Egypt, known since her death in 1955 as the Brooke Hospital for Animals, Cairo. All stories indeed may end in death, but perhaps not all deaths are equally tragic.[40]

Graphic Records

Horses had been integral to the prewar domestic and working lives of most civilians—rural and urban, agricultural and industrial—and the wartime efforts and fates of horses spent in the Great War went neither unnoticed nor unrepresented. Official reports provided dispassionate accounts of horses as matériel—as resources difficult and expensive to acquire, transport, and maintain, but also, in the end, expendable. Other media and genres offered more passionate and varied portrayals of them as, in effect, personnel. Recruitment and solicitation posters, for example, generally depicted horses as "dumb" but courageous comrades in arms, while photographs treated them either as valuable sources of labor or as valued fellow beings, and sometimes as both; paintings, public memorials, personal memoirs, and memorial essays typically portrayed horses as heroes, while literary works often used them as metaphors for human "wastage."[41]

Why not you?

The late nineteenth and early twentieth centuries were the golden age of the pictorial poster, and Great Britain and the United States used the medium extensively for military recruitment, particularly prior to the implementation of conscription. More rhetorical than mimetic in intent, recruitment posters trafficked in variants of one inspirational story: the military wants and needs men of strong mettle and in large numbers to ensure victory in its righteous cause. Not all recruitment posters featured horses, of course, but many used them as visual and thematic focal points. Varying in style of representation, and depending on the specific rhetorical purpose, posters might employ a naturalistic though highly romantic aesthetic to depict an individual soldier finding comfort and friendship in a horse, such as in an illustrated postcard (as opposed to poster) captioned PALS! (or COMRADES); or to portray a horse bearing a leader of soldiers on a national and spiritual crusade: PERSHING'S CRUSADERS, for example, superimposes a realistic image of Pershing over one of an incorporeal medieval crusader. Likewise, posters might use an exaggerated modern aesthetic to represent horses as embodiments of potency, or an embellished antiquarian style to represent them as icons of righteousness.

Two examples of those tendencies also serve to illustrate differences between American and British recruitment rhetoric. The American poster features a highly torqued image of a contemporary military rider, possibly suggesting Pershing, charging forward on an impossibly muscled gray saddle horse while contorting his body backward to point an accusing finger at the viewer. Neither the image nor the aggressive legend—*YOU* ARE WANTED BY U.S. ARMY—have borders separating rider and horse from viewer. The poster, in short, represents sheer power in order to threaten the viewer who does not sign up, while, more subtly, also promising the viewer who *does* sign up the chance to *become* powerful and threatening (Figure 1). By contrast, a British poster with the same goal of recruitment features an antiquarian image of a traditional St. George mounted on a charging white horse and slaying a dragon with his lance. A heavy border circumscribes the image, and lower and upper pediments carry the polite declarative: BRITAIN NEEDS YOU AT ONCE. The poster's stolid columnar composition and its stirring but bounded iconic image solicit the viewer with a historical—indeed, mythic—precedent for engagement in morally constrained martial violence justified by its righteous cause.

Posters using horses, of course, did so to attract humans—men in particular. Posters for cavalry and horse artillery enticed potential recruits with opportunities that ranged from getting into the fight as soon as possible, such as a poster promising "Quick Service Overseas," to gaining marketable experience for life after the war: one example, depicting a farrier shoeing a horse, reads, YOU CAN BECOME A Skilled Tradesman by JOINING THE ARMY. Posters promised polo players and other sportsmen opportunities to place their skills in national service, and others targeted viewers (presumably not sportsmen) to learn about horses and to work with them. Posters appealed positively to patriotism or duty (COME ON BOYS DO YOUR DUTY BY ENLISTING NOW!) or threatened negatively with guilt and shame—using the latter tactic, one poster depicted a trooper on his horse with the pronominally ambiguous caption "He's Ready to Fight, Are You?" while another depicted a canine in a Red Cross harness with the unambiguous caption EVEN A DOG ENLISTS—WHY NOT YOU? Generally, posters targeted to men emphasized bravado, adventure, and "manly" aggressiveness (ARTILLERY HEROES AT THE FRONT SAY "GET INTO A MAN'S UNIFORM"), while those to women, such as recruitment posters for the Women's Land Army, accentuated domestic animal husbandry or agrarian productivity (Figure 2).

While military posters used ethos to recruit human personnel, Blue Cross Fund posters used pathos to solicit funds for equine veterinary care. These posters also subtly equate horses and humans. One stunning example in bright colors represents a terrified horse going down behind yellow abstracted spears of light that simultaneously and paradoxically signify a shell blast, shield the blast's effects on the horse's lower body from the viewer, and symbolize the possibility of surviving those effects with proper medical treatment. A second example in black and white depicts a horse gone down, badly wounded from literally represented shelling and needing urgent medical help—a visual correlative for the poster's verbally expressed need to help "wounded" horses with "immediate" donations (Figure 3). Since the tack on both horses identifies them as saddle rather than draft or pack animals (or possibly as horse artillery "riders"), the posters tacitly draw attention to the absent men, by implication wounded or killed by the same shell blasts. Horses are the manifest visual focus of the posters, in short, while the equivalence of horses and humans as cannon fodder, intentionally or not, is their latent content.

Two other posters underscore that equivalence—positively and negatively—and also intentionally or not. A poster for Our Dumb Friends' League and the Blue Cross Fund features a draft horse in harness, the label on whose collar—"I Have done my Bit!"—identifies him as a war veteran, once presumably sick or wounded in battle but since saved and now performing productive civilian service. Expressly soliciting funds to rescue his "Comrades"—sick or wounded warhorses—the poster implicitly invokes the need to rescue his other comrades—sick or wounded soldiers—and to enable them likewise to perform useful civilian service. Finally, a Red Star Fund poster—captioned HELP THE HORSE TO SAVE THE SOLDIER over Fortunino Matania's widely circulated illustration, "Good-bye, Old Man"— depicts a clearly distraught soldier comforting a badly wounded and dying horse being abandoned on the road as the gunners and team move on. While the caption expressly asks us to help horses to save soldiers, the image implicitly asks us to help soldiers to save horses, an ambiguity underscored by the use of "old *man*" (my italics) to refer to the horse. Bearing witness to the mutual valor of horse and man, the poster also, intentionally or not, suggests their shared vulnerability in war to violent death.[42]

Apotheosis

If the recruitment posters tell one story, a small selection from the vast archive of photographs of horses in the Great War tells, or can be made to tell, two coexistent but competing stories. The first story, transparently sentimental, portrays horses as loved familial figures, often in the company of women or children, who as equine recruits in the British or American Expeditionary Forces shared human fellowship in the garrison or field, conveyed in images of individual soldier and horse pairs, while also satisfying their instinctual need for herd safety, conveyed in images of large numbers of horses traveling, resting, or eating together. Horses are shown in temporary shelters meant to provide camouflage from shelling and in appliances meant to protect against gas (Figure 4).[43] They are shown receiving first aid in the field, being transported to safety behind the lines by horse-drawn Red Cross equine ambulances, or undergoing surgery or recuperating from it in veterinary hospitals (Figure 5). Horses appear as courageous mounts negotiating extremely steep terrain, as versatile animals adapting to rivers and other obstacles, and, along with mules, as tireless workers carrying heavy packs (such as disassembled machine guns or multiple artillery shells) under horrible conditions.

The second story, grimmer and resolutely naturalistic, begins with horses from western Canada and the United States (and elsewhere) first herded to railheads and, after long rail travel to the coast, again herded onto ships for the transatlantic crossing, an ordeal that Timothy Findley has captured with harrowing detail in his novel *The Wars* (1977).[44] Unloaded in England by gangplank or sling, these North American equines were herded again onto railcars both to and from remount depots, were embarked and disembarked again in the Channel crossing, and then were transported again by train to their regimental assignments in France near the front. Once there, and the best human intentions notwithstanding, these horses and mules endured brutal working conditions in the mud and mire where, as noted earlier, they sometimes drowned or, if not able to be extricated, were shot (Figure 6). Loose and often wounded horses wandered the landscape, subject to severe privation, while thousands of the dead simply lay where they had fallen, their bodies bloated and waiting mass burial, as any number of appalling images document.

Those two stories may compete, but the photographs shaping them share three features. First, photographs of soldiers often picture only men, but those of horses almost always picture horses with men, emphasizing not only the value of horses to humans, but also the fellowship between humans and horses (and, again, sometimes their equivalent and shared vulnerability). Second, photographs of horses often depict them performing duties, or enduring hardships or deaths, that are similar—sometimes identical—to photographs of men in the same roles, implying the strong parallel or equivalent vulnerability between horses and humans in combat. And third, photographs invariably (because inherently) reflect the human *point of view* behind the camera, but they variably represent the human and equine *experience* as between two empowered agents, or between a human agent with power and an essentially powerless object, or between two objects, both of them lacking agency and both powerless to control their situations and fates. Humans, in short, may be able to master animals, but not necessarily other humans or themselves.

Those two stories, finally, can converge on grace notes that end, pace Hemingway, not in death but in its transcendence. Fairly uncommon, these images are formal and self-consciously "artful." Most photographs of trains of horses pulling guns or limbers, for example, are naturalistic and simply documentary, but one Australian image in silhouette is a visual elegy—a classical frieze conjuring a stately funeral cortege—whose harmonious composition and ideographic subject matter convey and universalize a harmony of nature, human, and horse that can transcend the squalor of war. Likewise, any of the hundreds of conventional images of mounted officers or troops, individually or in groups, crystallize, in effect, in a single stylized and thematically freighted image of a universal horse and rider, unified in silhouette as one iconic being and set against a sky symbolically cloudy but also brightening against the evidently lifting dark clouds (Figure 7). Photographs such as these represent an apotheosis in art of the thousands of horses and men who shared not only horror, but also heroism and the honoring of it.

Society of animal painters

In addition to graphic and photographic art, the Great War inspired a large body of fine art that represents the horse.[45] Done by traditional artists as well as the avant-garde, these paintings display styles and subjects ranging from panoramic battle scenes to formal portraits. Violent and visually vibrant impressionist renderings of open field actions, such as Thomas Dugdale's *Charge of the 2nd Lancers at El Afuli* (1918), for example, vie for attention with grim realist depictions of horse and foot soldiers entwined in grim and clotted close combat, such as Richard Caton Woodville's *Captain Francis Grenfell at Audregnies* (1914). Dynamic paintings of draft teams hauling guns and other weapons of destruction, such as Harold Power's *Bringing up the Guns* (1917) (Figure 8), foreshadow quieter images of the effects of those weapons and of other privations, such as Edwin Noble's less kinetic but far from static painting, *A Horse Ambulance Pulling a Sick Horse out of a Field* (n.d.). Equestrian portraits of generals, such as Jan van Chelminski's elegantly posed studies of Generals Haig, Pershing, Joffre, and Foch, may have less romantic élan than portraits of Napoleon or Washington made a century earlier, but they are no less idealized. And portraits of individual and specifically identified horses ennobled their subjects—as they always have done.

As John Fairley discusses in *Horses of the Great War: The Story in Art* (2016), the British and Imperial governments initially selected a small group of artists in an unofficial capacity, two years into the conflict, "to create a record of the war," and subsequently commissioned a sizable number of established painters as official war artists, dispatching both groups to Western and Eastern theaters.[46] Specialists in many different genres in their prewar careers, they included important sporting artists, such as Lionel Edwards and Alfred Munnings, both of them officers in the Remount Service, who, not surprisingly, gravitated toward warhorses as subjects. In one notable case, for example, *Hunting on the Salonika Front* (c. 1930), Edwards, a prolific illustrator as well as painter, simply transferred one of his favorite genres, the hunting scene, from the civilian to the military context. Munnings, arguably the most accomplished and versatile sporting artist of the period, similarly transferred his enormous skill in civilian equestrian portraiture to subjects such as Lord Mottistone (General Jack Seeley) on his celebrated charger Warrior. Munnings also reimagined foxhunters and horses going to or from meets as an orderly march of troopers and mounts (Figure 9); and he captured the vibrant action of his hunting and racing scenes in *Charge of Flowerdew's Squadron* (c. 1918), his celebrated depiction of a shock action en masse.[47]

As distinguished and accomplished as were Edwards and Munnings, Lucy Kemp-Welch (1869–1958) may be the most interesting of the war artists. The first president of the Society of Animal Painters, founded in 1914, Kemp-Welch was a prolific painter of draft horses, who made three signal contributions to Great War equine and equestrian art.[48] Commissioned by the British Parliamentary Recruiting Committee in 1914, she created the artwork for the iconic cavalry recruitment poster, FORWARD! Forward to Victory ENLIST NOW (1915) (Figure 10). Following later studies made at a Royal Field Artillery camp in 1916, Kemp-Welch executed *Forward the Guns* (1917), a well-received work—close in both subject and treatment to Power's *Bringing up the Guns*—exhibited at the Royal Academy in 1917 and purchased through the Chantrey Bequest Fund. Finally, commissioned in 1918 by the Women's Work Sub-Committee at the Imperial War Museum, Kemp-Welch produced two magnificent pieces commemorating the work of the many skilled civilian horsewomen who were recruited to manage remount depots and to train horses for military duty: *The Ladies' Army Remount Depot, Russley Park, Wiltshire, 1918* (1919), and *The Straw Ride—Russley Park Remount Depot, Wiltshire* (1919), each depicting groups of obviously skilled horsewomen exercising pairs of spirited horses, riding one while leading the other.[49]

Kemp-Welch's interest, in my view, derives not only from the exceedingly fine artistry of her work, and not even from her anomalous position as a woman among an almost exclusively male club of war artists—though both certainly command interest. Kemp-Welch, rather, and more broadly, can stand as an emblem for British feminism as the Great War dragged on and patriotism roiled the feminist nexus of pacifism, socialism, and suffragism, forcing women artists, activists, and intellectuals to navigate competing allegiances and alliances—a nexus explored at length by Adam Hochschild in *To End All Wars: A Story of Loyalty and Rebellion, 1914–1918* (2011). To take the examples at hand, one can read *Forward to Victory* and *Forward the Guns*, on the one hand, and *The Ladies' Army Remount Depot* and *The Straw Ride*, on the other, as opposed visions of male bravado and bellicosity versus female stolidity and pacifism; or, one can read them as corresponding visions of the equal equestrian expertise, and powerful sense of duty, shared by women and men in the war. In either case, Kemp-Welch's impressive scope speaks for itself.

Public memorials for British horses in the Great War, finally, are few but compelling. Early efforts to save and nurture former war horses, at once practical and memorial, include the Old War Horse Memorial Hospital founded by Dorothy Brooke in Cairo, and the Animal War Memorial Dispensary established by the RSPCA in London, both in the early 1930s—the latter in lieu of a sculptural memorial planned for Hyde Park.[50] Such a sculptural memorial, and there are two, would be long in coming. The Animals in War Memorial just outside Hyde Park, dedicated in 2004, commemorating all animals "that served and died alongside British and Allied forces and in wars and campaigns throughout time," is dominated by bronzes of a horse and two mules. It is inscribed simply, "They had no choice." The War Horse Memorial, near Ascot, followed in 2018. Dedicated to "war horses that lost their lives during World War One," it comprises a bronze of a single mournful and emaciated horse. Most poignant, however, may be the memorial erected at Saint-Jude-on-a-Hill church, in London, in 1926. Its statue was stolen, but its surviving bronze plaque speaks with eloquence: "In grateful and reverent memory of the empire's horses who fell in the Great War (1914–18). Most obediently, and often most painfully, they died."[51]

Literary Reckonings

Government planners, analysts, and accountants, as noted earlier, treated horses primarily as resources or matériel—invaluable but ultimately expendable sources of power—while essayists and memoirists, particularly cavalrymen, treated them primarily as personnel, as individuals or even personalities—invaluable and irreplaceable comrades-in-arms. Journeymen authors tended to portray warhorses as equine heroes, while more notable literary figures employed them and their fates as metaphors for human beings, human carnage, and human moral ambiguity. In almost all cases, however, writers invoked either or both of two related themes: warhorses could not speak, and warhorses had no choice.

Dumb colleagues

Those themes emerged when Victorian sensibilities met modern warfare. In "Dumb Colleagues," for example, a "letter" from the Boer War in *A Subaltern's Letters to His Wife* (1901), Lieutenant Reginald Rankin (1871–1931) turns period anthropomorphism on its head, arguing that the alleged "brute-nature" of the horse does not preclude bravery, because "The courage of the soldier is animal courage. There is . . . no qualitative difference between a man's and a horse's courage." Anticipating the theme common in Great War recruitment posters, Rankin contends that the difference lies in neither character nor acts, but rather in speech: "Truly, if these gallant creatures have not their reward," in Rankin's inversion of values, "it is because they cannot ask for it. Men expose their stupidity in speech, and animals veil their intelligence in dumbness." Despite his bold conceit, his grim litany of equine suffering, and his central point—"Incalculably greater than the sufferings of the men, great as they were, were the sufferings of the horses and mules"—Rankin offers another apotheosis: three horse carcasses "piled one upon another" resolve into the image of a marble sculpture of "three friends who, in their death, would not be divided."[52]

Two decades and a cataclysm later, three eloquent custodians of equine valor in past and future wars used a more modern rhetoric to acknowledge the mass suffering and wastage of horses in the Great War while also celebrating them as heroes. In *The Horse and The War* (1918), written while "still in the midst of the raging tumult," Sidney Galtrey (1878–1935), captain, Royal Army Service Corps, told a story whose protagonists are the warhorse and mule in their collective numbers, not as identified individuals.[53] Many postwar years later, in *"The Horse in War" and Famous Canadian War Horses* (1932), Lieutenant-Colonel D. S. Tamblyn (1881–1943) offered a compendium of anecdotes about battles, actions, horses, and men—often anonymous—followed by biographies and portraits of individual celebrated warhorses. And shortly after Tamblyn, Lord Mottistone (General Jack Seely) (1868–1947) published *My*

Horse Warrior (1934), a book-length biographical memoir of a single, and singular, decorated equine hero and veteran.

Galtrey's *The Horse and The War*, illustrated by Lionel Edwards, a tribute to "the hundreds of thousands of horses and mules that have been gallantly aiding the Empire's Cause," melds the political and memorial. A prefatory note from Field Marshal Haig, praising "the wisdom of breeding animals for the two military virtues of hardiness and activity," and an epigraph quoted from a letter from H. M. the King to Haig, expressing his "pleasure" that "the new Armies fully uphold our national reputation as good horse-masters," foreshadow Galtrey's advocacy for a postwar national breeding program and his praise of the Remount, Artillery, and Veterinary Services for their excellent horse management and care—a sore point, His Majesty notwithstanding, since the Boer War. Galtrey's political agenda detracts nothing, however, from his heartfelt challenge to the reader that "when, either now or years hence, you come to read of the defeat of great German Armies in their plans to crush and batter the British out of existence, you will perhaps spare a grateful thought for the horses and mules which in their thousands made our salvation possible."[54]

Addressing "the people of this country, of our Empire, and of the countries of our Allies," Galtrey specifically wants to debunk the alleged "fact that motor haulage has largely displaced horses [in war]" and to demonstrate how in the Battle of the Marne, for example, "the horse and the mule were essential for the guns, the transport, the ammunition columns, and all arms of mounted troops." Employing the simple but effective narrative strategy of guiding the reader through the stages of a warhorse's career from breeding and purchase, through transport and active service, to casting in its various forms, Galtrey focuses on the light draft horses and mules imported from "America" (his umbrella term for the United States and Canada): the light draft horse is "the outstanding success of the war," "the war-horse that has made history," "the real equine hero of the war," "the real horse of the war"; and the mule is "the other conspicuous success," "the most serviceable and satisfactory animal used in the war," "perhaps less appreciated" but "just as essential [as the light draft horse] in his own peculiar way." The mule earns a dedicated chapter, as does the light draft, the latter chapter being Galtrey's call for a state breeding program for Percheron crosses.[55]

D. S. Tamblyn's *"The Horse in War" and Famous Canadian War Horses* is an even more eulogistic valedictory to "those who cannot speak for themselves"—a theme of "dumbness" that serves not only as the book's leitmotif but also as its raison d'être: Tamblyn will speak now for the horses who could not speak then for themselves.[56] *"The Horse in War,"* though, despite its title, actually speaks more about men than horses, and it treats horses, mainly artillery horses, often en masse and anonymous, as courageous, primarily, in what seems an essentially *passive endurance* of suffering. Consequently, Tamblyn catalogues in grisly detail many types and instances of that suffering—such as razor bombs designed to disembowel horses—and he plumbs the moral demands that such suffering placed on soldiers.[57] Attempting to extricate horses or mules sunk "to their knees in liquid mud," for example, risked the lives of three or more others in the team, but leaving them resulted in their likely "becoming wounded through shell fire and lying in agony for hours," so shooting them was the difficult but also the only humane option.[58]

As a eulogist, Tamblyn seeks to redeem equine suffering and human responsibility for it—as well as challenge one type of human profiteering from it—asserting that, after all, it was "the horses [who] bore their riders to fame." Consequently, in the second half of his compound book, he reassigns fame from officers to their horses and focuses on the *active hero-ism*, as opposed to passive endurance, of those horses. Following a long tradition of celebrating warhorses that had extended from Alexander's Bucephalus, through Wellington's Copenhagen and Napoleon's Marengo, to Grant's Cincinnati and Lee's Traveller, and contemporary peers, *Famous Canadian War Horses* devotes nearly fifty pages to encomia for individual warhorses, all of them officers' chargers. General Sir A. W. Currie's gelding Brocklebank, in a manner of speaking, leads this "herd" of some twenty horses and mares: "No horse was more outstanding in the Great War than this animal." Tamblyn's treatment of identified cavalry chargers rather than anonymous artillery horses may reflect pieties of rank and breed, but it also accords those individual horses an equine identity that transcends sentimentalism and anthropomorphism alike.[59]

My Horse Warrior, Lord Mottistone's homage to its titular figure, illus-trated by Alfred Munnings, exalts those pieties. A memoir of Mottistone's twenty-six years with his thoroughbred charger, Warrior, as "constant com-rades and friends," *My Horse Warrior* mainly focuses on their four years of service during the Great War. Despite setbacks, such as the retreat following the Battle of Passchendaele in 1917, their service culminated in a spirited

cavalry charge in the Battle of Moreuil Wood in March 1918, led by Mottistone (then General Jack Seely) and Warrior—a charge decisive in the outcome of the war. Mottistone describes the moment repeatedly as not only
Warrior's (and his) "greatest ordeal," but also, and more grandly, as "that
great day when [Warrior] had galloped through the British and German
front lines to save Amiens and the Allied Cause." Shamelessly hagiographic,
My Horse Warrior attributes to Warrior virtually all equine and human virtues and no physical, mental, or moral vices. It also presents him as a charismatic leader both beloved by troopers and celebrated by civilians.[60]

We can take Mottistone at his word when he says that he wrote his
book from "a sense of duty" to "my faithful friend," though the book suggests other motives as well. Mottistone uses Warrior's history of care, for
example, to exemplify and promote kind treatment of horses. He defends
the position that "the horse is vital to man in modern war"; he debunks
the belief "that all battles will be conducted by mechanical means"; and he
urges "an adequate supply of horses in time of war." Less obviously, almost
surreptitiously, he proffers a counterpoint to the view of the war as a "big
lie" that resulted in a "lost generation," a view that had gained considerable traction since the late 1920s. His many references to the bravery and
tenacity of Germans in the trenches, for example, could be read simply as
grace notes, instances of aristocratic military etiquette, were it not for
Mottistone's controversial trips to Nazi Germany in 1933 and 1935, and
his vocal defense in Parliament, throughout the 1930s, of the policy of
appeasement. Finally, Mottistone is either unconsciously reflecting, or,
more likely, consciously affirming, the nostalgic interwar English zeitgeist
and its multiple elegies for a dying empire.[61]

By way of contrast, Private James Robert Johnston's *Riding into War:
The Memoir of a Horse Transport Driver, 1916–1919* (2004) is based on
"Memories of the Great War," a personal memoir written in 1964 by an
unknown enlisted man for purposes of private remembrance, rather than
by a decorated officer for purposes of public acclaim. In it, Johnston
(1897–1976) tells the story of his service on the Western Front in the 14th
Canadian Machine Gun Company, a unit outfitted with fifty-four horses
for its "horse-drawn limbers and wagons" and as pack animals.[62] Johnston
treats those horses with matter-of-fact respect and without sentiment—as
neither resources nor heroes, but rather as reliable, tireless workers in
extremely hazardous circumstances, including and especially water-filled

shell holes. "The trouble with transport work in a war," he writes with reference to both men and horses, "is that while under fire, one has to stay above ground and has no protection from shelling or machine gun fire, whereas foot troops have a chance of protection from a shell hole, trench, etc."[63] Far from being a titled officer glorifying chargers, in short, Johnston is a working-class recruit unimpressed with command, and dedicated to honoring humans and draft horses as both laborers and victims.

Hemingway's horses

Unlike nonfictional memoirs, imaginative literature of the Great War paid horses scant attention—a notable exception being Siegfried Sassoon's fictional trilogy, *The Memoirs of George Sherston* (1937). Extending a tradition that treats foxhunting as training for war, its first two volumes—*Memoirs of a Fox-Hunting Man* (1928) and *Memoirs of an Infantry Officer* (1932)— juxtapose the openness and mobility of the chase with the claustrophobia and deadlock of the trenches: foxhunting, Sassoon implies, could prepare no one for *this* war. As those companion titles suggest, Sassoon's story concerns the man more than his horses, though Sassoon (or, more properly, his narrator, Sherston) frequently invokes by name his field hunters in prewar England, and he fondly remembers stolen moments of "canter[ing] about the open country by myself" on military horses while in service in France.[64] Sassoon also advances three familiar themes: artillery and transport horses died in the thousands; their corpses littered the landscape; and they had neither voice nor choice in the matter.

Sassoon's contemporary, Ernest Hemingway (1899–1961) served as a driver for the American Volunteer Motor-Ambulance Corps in France, the experience behind his three important works of Great War fiction: *In Our Time* (1925), *The Sun Also Rises* (1928), and *A Farewell to Arms* (1929). None featured horses. In 1930, though, Hemingway added an introduction, "On the Quai at Smyrna," to a new edition of *In Our Time*, employing his signature flat prose to invoke a grotesque event, in this case the slaughter of horses and mules in 1922 in an evacuation following the Greco-Turkish War: "they just broke their forelegs and dumped them into the shallow water."[65] The incident provided Hemingway a metaphor for human slaughter in the Great War that he reused twice, verbatim, in *Death in the Afternoon* (1932), his "introduction to the modern Spanish bullfight."[66] A subtle and generically heterogeneous book of over five hundred pages, and ostensibly about neither horses nor the Great War, *Death in the Afternoon* offers a profound meditation, albeit metaphoric, on horses *in* the Great War.

Hemingway constructs that meditation on one powerful insight: The bullfight is not sport. It is tragedy. As Hemingway explains, "The bullfight is not a sport in the Anglo-Saxon sense of the word. . . . Rather it is a tragedy," played out ritually by man and bull, ending in "certain death for the animal."[67] The horse, who plays a tertiary role in the performance of the tragedy, plays the primary role in the theme that informs *Death in the Afternoon*: the spectator's reaction to the horse's wounding or death in the performance, a theme launched by Hemingway in his opening sentence.[68] The theme has two main components: One, as an aficionado, Hemingway *must* acknowledge the goring of the horse as inherent, if incidental, in the tragedy's performance; and, as a spectator, though one who does "not feel any horror or disgust whatever at what happens to the horses," he *must* accept his complicity in the goring if he is to honor the aesthetic and moral integrity of the ritual and remain a "true-story teller."[69] Two, civic officials who mandate the "peto," a "quilted mattress" used to protect the horse's abdomen, cause fewer horses to be killed, but nearly all horses to be repeatedly wounded; as advocates of "something designed to allow the horses to suffer while their suffering is spared the spectator," the officials dishonor the ritual, as do the spectators who welcome being spared.[70]

Hemingway is shaping an allegory for the war based on the bullfight by conjoining, one, his belief that "the only place where you could see life and death, i.e., violent death now that the wars were over, was in the bull ring"; with two, his insight that bull fighting is not sport but ritual tragedy; and three, his indictment of all those who dishonor the ritual—in addition to spectators, these include a triumvirate of profiteers: the purveyors of horses, the promoters of bullfights, and the civic authorities charged with oversight.[71] Combatants in the ritual of the war—men and horses—suffered the violent wounding and death inherent in it. Government officials and war profiteers, who enabled and benefited from the violence while sparing civilians as much as possible from "seeing" its effects, robbed the war of its meaning (or, more cynically, disguised its meaninglessness) and dishonored the ritual of combat. And the civilian "spectators" who cheered both the fighting and the victory from afar, as a result, also dishonored both the ritual and their own integrity. Ultimately, though, if the abstracted ritual of violent death in the bullring can serve as allegory for the conduct of the war, it can serve as nothing but allegory for "life and death, i.e., violent death," as actually suffered in the war.

Death in the Afternoon, in sum, offers not only a meditation on horses in the war but also an extended metaphor for the war's carnage and for a generation's shattered values. Hemingway distrusted grand metaphors, however, and famously wrote in *A Farewell to Arms*, "Abstract words such as glory, honor, courage, or hallow were obscene beside the concrete names of villages, the numbers of roads, the names of rivers, the numbers of regiments and the dates."[72] Although he risked metaphoric abstraction in *Death in the Afternoon* by using the bullring to stand for the war and horses to stand for humans, he also mitigated the second abstraction by using horses in the bullring to stand for horses in the war—the same animal with the same fate. In so doing, he accorded warhorses concrete dignity as animals maimed and killed in a mass ritual tragedy where their role was tertiary, their fate inherent, and their purveyors and witnesses without honor. Slaughtered by the millions in that tragedy, to quote again the memorial at Saint-Jude-on-a-Hill, "most obediently, and often most painfully, they died."

In "Some Reactions of a Few Individuals to the Integral Spanish Bullfight," part of *Death in the Afternoon*, Hemingway profiles a dozen men and women, using a variety of identifying features, including whether or not the individual is a horseman or horsewoman. Though he recalls "an Englishwoman who . . . was so overcome by the horses being charged by the bulls that she cried as though they were her own horses or her own children who were being gored," his profiles overall do not indicate, or even imply, whether or not horsemanship determines reaction to the bullfight and the fate of the horse in it—and by metaphoric extension, presumably, reaction to the war and the fate of the horse in it. The profiles admit no inference. Since civilian horsemanship in England essentially meant foxhunting, however, a report in the *Manchester Courier* from 1915 may be relevant: "The Masters have gone to war; the Hunting Men, the officers, have gone to war; the Hunt servants and stablemen have gone to the war, and above all the horses have gone." Many of the huntsmen, and even more of their horses, did not come back. For that and other reasons, writers mined foxhunting, like bullfighting, for metaphors of war.[73]

Figure 1. U.S. Army. Recruitment poster.

Figure 2. National Service Women's Land Army. Recruitment poster.

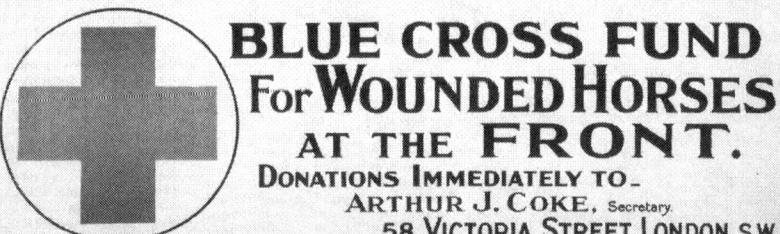

Figure 3. "Our Dumb Friends' League." Blue Cross Fund for Wounded Horses at the Front. Solicitation poster.

Figure 4. Army Efficiency Tests. Horses and Men in Gas Masks. September 1917.

Figure 5. Dressing Station. Flanders front. 1917.

Figure 6. British Horse Sunk in the Mud. Battle of Ypres. 1917 or 1918.

Figure 7. British Cavalry Patrol. 1917.

Figure 8. Harold Power. *Bringing Up the Guns.* 1917.

Figure 9. Alfred Munnings. *Lord Strathcona's Horse on the March.* 1918.

Figure 10. Lucy Kemp-Welsh. Illustration for *Forward! Forward to Victory Enlist Now*. Recruitment poster. 1915.

5

Hunting in the Trenches

I think we are in rats' alley
Where the dead men lost their bones.

T. S. Eliot
The Waste Land (1922)

British military writers in the eighteenth and nineteenth centuries frequently advocated "riding to hounds" as fit preparation not only for cavalry
service but also for soldiering in general—an idea that retained its vitality
in and after the Boer War. Erskine Childers, to cite just one example
among many, asked in 1910, "Was not our own possession of sporting and
hunting aptitudes, embodied in Englishmen and Colonials alike, our very
salvation in South Africa?" And answered, "Of course it was."[1] Though the
idea of foxhunting as military preparation, primarily for officers, declined
as a literal proposition after the Great War, it did not disappear. It lived on,
rather, as a figurative sentiment—one that still retains purchase on the
British imagination. Siegfried Sassoon's semiautobiographical trilogy, *The
Memoirs of George Sherston* (1937), stands on both the prewar history of the
hunting creed and its postwar revival as a nostalgic trope.

The first of the Sherston trilogy's three volumes, *Memoirs of a Fox-
Hunting Man* (1928), traces Sassoon's eponymous alter ego from hunting
as a boy and youth on an idyllic English landscape to killing as a young
man on a hellish Western Front. The volume implies that foxhunting, military tradition notwithstanding, could have prepared no one for the
uniquely modern horrors of the Great War. Yet its successor volume and
the trilogy's centerpiece, *Memoirs of an Infantry Officer* (1930), hints that
hunting was precisely what equipped Sherston for his battlefield heroics,

135

and the final volume, *Sherston's Progress* (1936), for his subsequent moral "progress" to antiwar activism. Though Sassoon's own autobiographical memoirs posit and explore the same thematic nexus of foxhunting and warfare, the memoirs of his character Sherston do so in a more patently "literary" narrative form.

Something Like a Life Work

Though primarily known as a Great War poet, Sassoon (1886–1967) joins Robert Graves and Edmund Blunden as one of the most important British Great War memoirists to have served on the Western Front: Blunden's *Undertones of War* (1928), Graves's *Goodbye to All That* (1929), and Sassoon's *The Memoirs of George Sherston*, together with Vera Brittain's home front memoir, *Testament of Youth* (1933), dominate the genre. That stature is no small matter, given that eminent British literary and military figures, as well as veterans and observers of lesser or no renown, produced scores— really, hundreds—of Great War memoirs ranging from best sellers with multiple editions in the thousands to privately printed recollections with a single edition in the tens (for bibliographies of the latter, see Lengel, *World War 1 Memories*, and Donovan, *In Memoriam*). Sassoon, however, exceeds even Graves and Blunden in one respect: the grip that his idée fixe had upon him.

As Paul Fussell observed in his seminal study, *The Great War and Modern Memory* (1975), that is, "remembering the war became something like a life work" for Sassoon as diarist and memoirist. Sassoon kept extensive diaries during and immediately after the war that served as a basis for the three volumes of semi-fictional memoirs forming the Sherston trilogy, as well as three volumes of nonfictional memoirs covering a broader span.[2] The Sherston volumes cover its protagonist's life (and, transparently, Sassoon's) from roughly 1895 to 1918. Published only two years apart, *Memoirs of a Fox-Hunting Man* and *Memoirs of an Infantry Officer*, as their consonant titles suggest, are a linked pair. The first has become a celebrated classic in the literature of foxhunting, and the second an exemplar of Great War frontline writing. Published four years later, the third volume, *Sherston's Progress*, is a desultory affair, much shorter than its predecessors and, for the most part, a barely enhanced transcription of Sassoon's diaries. Taking the trilogy's form and themes to their conclusion, however, it also provides Sherston a vantage from which he can reflect not only on the events experienced and narrated in the three volumes, but also on his recollection and narration of them.

Despite the falling off in literary quality in its third volume, the trilogy coheres throughout as a story. Tracking Sherston's evolution as a young rural sportsman to his installation as a newly minted officer in 1914, *Fox-Hunting Man* concludes with his waking from an idyllic dream of the hunt to a grim view over no-man's-land. *Infantry Officer* recounts Sherston's life at the Western Front from 1915 to 1917, including, one, the reckless single-handed attack on a German trench that earns him a Military Cross and the sobriquet "Mad Jack"; two, the growing "anti-war bitterness" that issues in his public antiwar statement and activism and, as a consequence, the army's retaliatory diagnosis of him as "shell-shocked"; and three, the ensuing hospital confinement where "this [second] volume can conveniently be concluded." *Sherston's Progress* then plumbs his psychiatric therapy, the internal conflict resulting in his return to action in 1918, and his sustaining the wound that "ended my last week at the War. And there, perhaps, my narrative also should end." And there, indeed, it does.[3]

More complicated in structure than in story, the trilogy views and presents its world through two almost inverse perspectives. On the one hand, it offers a Great War commentary, or, more precisely, an antiwar indictment, with a tale of individual redemption as its framework, while, on the other hand, it unfolds a conversion narrative, or a tale of personal redemption, with the Great War as its backdrop. Though the former perspective is generally more explicit than the latter, the trilogy constantly toggles between them, shifting foreground and background throughout its length and sustaining them in a dynamic equipoise. Though Fussell identifies this as a "remorselessly binary" vision in Sassoon's poetry and prose, one defined by extreme oppositions, "flagrant dichotomies," "insistent polarities," and "exaggerated antitheses," the vision seems more bifocal than binary, comprising a dynamic interplay, rather than a static opposition, between perspectives. This holds even for *Infantry Officer*, with its steadier focus on the nature and horror of the war itself.[4]

A crime against humanity

The Sherston trilogy, and *Infantry Officer* in particular, provide a litany of the staple images, motifs, and themes found in the large body of Great

War poetry and prose, mainly from the 1920s, that refused either to glorify England's declaration or prosecution of the war or to celebrate its victory:

The presumptive holy war to save civilization instead turned France into a "Stygian" landscape haunted by "an army of ghosts," "a dreadful place, a place of horror and desolation which no imagination could have invented" and where Sherston stumbles upon a disembodied hand in the mire, one that "seemed to be pointing at the sky with an accusing gesture." What "the blasted Bishops call the Great Adventure" could lead even Sherston to "working myself up into a similar mental condition, as though going over the top were a species of religious experience." Pieties were rampant: "And those who are killed in the War—they help us from up there. They are helping us to win."[5]

The war effort that initially depended on falsely "optimistic news" about the British effort and false reports of German atrocities eventually demanded manufacture of a "camouflage War . . . to aid the imaginations of people who had never seen the real thing"; in "War's demented language . . . lies were at a premium," and "any lie is a good lie as long as it stimulates unreasoning hatred of the enemy." As Graves writes in *Goodbye to All That*: "The civilians talked a foreign language; and it was newspaper language." To have disbelieved it, Sherston adds, "would have been unpatriotic."[6]

The "amateurish mismanagement and incompetence of British military leaders," Sassoon wrote in his later memoir, *Siegfried's Journey* (1946), made fodder of brilliant young officers and "victims" of ordinary soldiers. Civilians, not much better than military brass, fell into one of three groups: gulled citizens, self-deluded aristocrats, and "gross profiteers" gorging on the carrion of soldiers, as symbolized by their gluttony in the safety of London restaurants: "They ordered lobsters and selected colossal cigars," Sherston recalls.[7]

The volunteers who enlisted with patriotic zeal and youthful adventurism in August 1914, expecting to be home for Christmas, soon were put straight by Kitchener's prediction "that it would be three years" before their return. They eventually came to see themselves in an "endless war." Sherston imagines, invoking modern suburbia, "that if the war lasted a few more years we should be coming to the trenches every day by train like city men going to the office." Sassoon recalled of November 1918, "Even now it wasn't easy to absorb the idea of the War being over."[8]

The military strategists at the beginning of the twentieth century who foresaw the next war as both "total" in its mobilization of men and resources and "modern" in its mechanization of combat and combatants turned out to be right. Sherston finds himself in "the Machine"—"the Sausage Machine" that turns men into meat in order to manufacture "Military History." As Liddell Hart wrote in 1928, "Thus mechanical butchery became the essence of war, and . . . attained its tragi-comic climax on the Western front in the World War."[9]

Humans devolved: vast scale rendered human enterprise "ant-like activity"; trench warfare reduced human beings to creatures either crawling in the mud or moving "mostly on all fours"; and attrition resulted in conscription of men no better than beasts of burden, "mostly undersized, dull-witted, and barely capable of carrying the heavy weight of their equipment."[10]

The war was a ubiquitous, inescapable reality for combatants both on the front and convalescing at home. Sherston realizes during his first convalescence that "I couldn't be free from the War; even this hospital ward was full of it"; and, likewise, during his second convalescence, "I felt a sudden sense of the unreality of my surroundings. Reality was on the other side of the Channel, surely." Making a separate peace is impossible: "Going back to the War as soon as possible was my only chance of peace."[11]

In a similar paradox, men at the front, including Sherston, came to see dying in battle ("going west") as an inevitability to be dreaded but, simultaneously and paradoxically, embraced: "For me, the idea of death made everything seem vivid and valuable." While officers might view "the idea of the supreme sacrifice" differently than recruits, combatants in an impersonal war based on long-distance shooting and shelling, in the end, are killed almost with the randomness of a lottery, regardless of rank or virtue. Hemingway built a literary corpus and career on those insights.[12]

The horror of trench warfare and the inadequacy of language to express it—especially language debased by lies and propaganda—left veterans isolated but also insulated. As Sherston succinctly puts it: "We were the survivors; few among us would ever tell the truth to our friends and relations in England. We were carrying something in our heads which belonged to us alone, and to those we had left behind us in the battle." As Sassoon writes in *Siegfried's Journey*, "The demobilized young felt that it was no use attempting to explain to their elders the realities of an experience from which they had been excluded, or that those elders had any right to express opinions about it."[13]

The war resulted in a widely memorialized if comparatively small missing or "lost" generation of ruling class, public school and university educated young officers, and a far larger, and less commemorated, population of "ordinary soldiers." In a comparative study of Great War literature, *The Generation of 1914* (1979), Robert Wohl identified and unpacked "the narrative pattern that characterizes . . . the late 1920s and early 1930s: innocence, followed by betrayal and defeat at the hands of the older generation." No exception, Sherston exits his narrative with "no conviction about anything except that the War was a dirty trick which had been played on me and my generation."[14]

With that bill of particulars, Sherston indicts the perpetrators of "the War, which, as everyone now agrees, was a crime against humanity."[15]

On the way to redemption

Those perpetrators included governments and military leaders, churches and religious leaders, industries and war profiteers, and a gullible public of noncombatants. It follows, however, that officers and soldiers who did the fighting—whether voluntarily or involuntarily, willingly or unwillingly—were enablers as well as victims, though Sherston takes pains, not always convincingly, to exonerate "ordinary soldiers." While Great War poetry and prose (including Sassoon's), moreover, explicitly leaned toward the secular rather than the sacred, they also represented the "crime" as immoral and so, implicitly, a sin. While the war's perpetrators were guilty of a crime demanding punishment and reparation, its enablers bore the guilt of a sin wanting atonement and redemption, or, as Sassoon calls it in *Siegfried's Journey*, "absolution."[16]

Fussell's overall assessment of the Sherston trilogy as "elaborately structured to enact the ironic redemption of a shallow fox-hunting man by terrible events" aligns with that proposition. Sherston the sportsman-turned-soldier-turned-witness had joined the war effort voluntarily and willingly—indeed, enthusiastically—but he eventually "lost my faith" in the war's righteousness and came to believe "that if the War were to end tomorrow I should be starting on a new life's journey in which point-to-point races and cricket matches would no longer be supremely important," and in which he would endeavor to "take some small share in the real work of the world." What such "real work" might be, however, and whether it would be "good work," remain key questions explored by the trilogy.[17]

Sherston's main literary allusions provide some answers, as does his progress from antiquarian book collecting to studious reading, one that closely parallels in a minor key his progress from sportsman to activist.[18] As novice sportsman, Sherston "knew I ought to read *Paradise Lost* and *The Pilgrim's Progress*," but had no time for "such edifications." As an older memoirist, though, he evokes Milton and *Paradise Lost* to underscore not only his and his generation's "fall" into knowledge, but also his realization that the only paradise available to them consists in *remembrance* of an ideal lost paradise represented by English countryside and country life. More important, he not only invokes Bunyan and *Pilgrim's Progress*, the exemplary British parable of salvation, in the title and epigraph of his concluding volume and in ubiquitous references throughout the trilogy to his "pilgrimage," his encounters with "Despair," and his moral "progress," but he also borrows the structural and thematic foundation of his *Memoirs* from Bunyan.[19]

Sherston's tour through an infernal war and purgatorial convalescence, and even to a lost paradise, in addition, inevitably recalls Dante's journey in the *Divina Commedia* through the underworld of the *inferno* and *purgatorio* to the *paradiso* and Beatrice's ideal love, while references to "the new life toward which [Sherston's psychologist Rivers] had shown me the way" recall Dante's *La Vita Nuova* and its celebration of earthly love as both synecdoche for divine love and vehicle for achieving it.[20] The trope that had animated that greatest of Great War poems, T. S. Eliot's *The Waste Land* (1922)—the trope that Denis de Rougemont soon would trace through history in the definitive *Love in the Western World* (1939)—also animates the celebration of comradery and homoeroticism ubiquitous in Great War literature and pervasive in the Sherston trilogy.

Sherston's (and Sassoon's) redemption, then, entails first experiencing a fall into the knowledge that paradise has been lost, then journeying from "sin" through suffering to atonement, and finally embracing earthly (and possibly divine) love. Redemption also entails Sherston's bearing witness to what he has seen, experienced, and learned in the war, not only in the private service of finding a peace for himself, but also in the public service of doing the "good works" of creating a memorial to the dead and a cautionary tale for the living—notwithstanding how "difficult it is to recover the details of war experience," or even to believe "that I was there at all." As participant and narrator, "I wanted to know—to understand—before it was too late, whether there was any meaning in this human tragedy which sprawled across France." Sharing what he came to know becomes his duty and his task.[21]

The complex advantage of being a soldier poet

The Great War "usually enters our minds not as history, but as literature," Richard Holmes observes in *Tommy: The British Soldier on the Western Front* (2004), raising the question of "literariness" (to use a clumsy term) that overarches Sherston's trilogy and that Wohl explored in *The Generation of 1914.* The products of well-bred families and elite schools, Wohl argues, the young men who served in numbers as subalterns in the British officer corps "had all been trained in a literary tradition that translated quotidian and unpleasant reality into elevated sentiment and diction. . . . They arrived in France with a ready-made store of images and metaphors with which to interpret their experience." While the lesser poets among them failed to transcend those limitations, the best created "a new poetry, and ultimately a new literature, which represented the fate of the English generation of 1914 in radically different terms." Sassoon was its "leading exponent."[22]

Sassoon describes his progress, in *Siegfried's Journey*, from writing poems influenced by Rupert Brooke and expressing "the typical self-glorifying feelings of a young man about to go to the Front for the first time," to writing "a few genuine trench poems," in early 1916, "dictated by my resolve to record my surroundings, and usually based on the notes I was making whenever I could do so with detachment. These poems aimed at impersonal description of front-line conditions, and could at least claim to be the first things of their kind." More thoroughly disillusioned by

December 1916, "I had one compelling mental objective to support me. This was the need to obtain further material for war poems. My strength of mind thus consisted mainly in a ferocious and defiant resolve to tell the truth about the War in every possible way." Sassoon's lecture tour in the United States, in early 1920, undertaken "with the purpose of revealing the realities of front-line soldiering," displays the same resolve, as do the first two volumes of the Sherston trilogy, albeit with formal and focal differences and the distance of a decade.[23]

When Wohl argues that Blunden's *Undertones of War* (and Sassoon's *Fox-Hunting Man*) "anticipated and helped to shape the approach of later war books by abandoning any attempt to describe the general context of warfare within which his personal experience took place," he raises what is only a superficial paradox. Sherston cautions his readers, "It is my own story that I am trying to tell, and as such it must be received; those who expect a universalization of the Great War must look for it elsewhere. Here they will only find an attempt to show its effect on a somewhat solitary-minded young man"; he claims that "I am no believer in wild denunciations of the War; I am merely describing my own experiences of it"; and to a pacifist editor's request for "something outspoken so as to let people at home know what the War was really like. I offered to provide such details as I knew from *personal experience*" (my italics). Sassoon shared with Hemingway that commitment to the concrete over the abstract and the specific over the general, though he realized it in a very different prose style.[24]

Sherston intends his memoirs, in other words, as Sassoon did his poems, to record surroundings and describe frontline conditions impersonally in order to demonstrate their effects on an individual, paradoxically, through his personal experiences. The memoirs thus emphasize external rather than internal reality, bodies and psyches rather than physiology or psychology, because "the intimate mental history of any man who went to the War would make unheroic reading," and because, he says in *Infantry Officer* with obvious irony, "I am beginning to feel that a man can write too much about his own feelings, even when 'what he felt like' is the nucleus of his narrative." It is only much later, and in the different order of reality narrated in *Siegfried's Journey*, that Sassoon finds, while writing *Sherston's Progress*, that, "I am now more inclined to analyse and investigate the inner history of a course of action that I have never regretted and for which there was no apparent alternative."[25]

The problem, of course, is that Sherston's memoirs are not only memoirs, and thus intrinsically internal and subjective, but are also highly self-conscious as *being* memoirs. They allude, for example, to the most monumental of early twentieth-century fictional memoirs, Marcel Proust's *A la recherche du temps perdu*, whose celebrated English translator, C. K. Scott Moncrieff, was an acquaintance of Sassoon's. Sherston, indeed, explicitly addresses the problems of lost time—"time seems to have obliterated the laughter of the war. I cannot hear it in my head"—and of narrative framing—"I am always reminding myself to be ultra-careful to keep my story 'well inside the frame.' But I begin to feel as if I were inside the frame myself." More important, when Sherston indicates that the purpose of his memoirs is "to show myself as I am now in relation to what I was during the War," he points to issues inherent in all retrospective first-person narratives and particularly keen in a fictional memoir whose author would claim it, albeit coyly, to be his own nonfictional memoir as well.[26]

All retrospective first-person narratives, that is, have an author, narrator, and central character with various kinds and degrees of ironic distance between them. The Sherston trilogy relies on a long ironic distance between what Sherston as character saw or failed to see and what Sherston as narrator sees, together with a shorter distance between what Sherston the narrator sees or fails to see and what Sassoon the author sees. Those ironies are complicated, moreover, by Sherston's at once being and not being Sassoon: Sassoon writes in *Siegfried's Journey*, "My [Sassoon's] experiences during the next three weeks . . . have already been related in *Memoirs of an Infantry Officer*"; and "His [Sherston's] experiences were mine, so I am spared the effort of describing them." The crucial difference is that "Sherston was a simplified version of my 'outdoor self.' He was denied the complex advantage of being a soldier poet." That distance informs *Fox-Hunting Man* and, to a lesser degree, *Infantry Officer*.[27]

From Fox-Hunting to Boche-Hunting

Sherston's trilogy, put differently, belongs to the genre bildungsroman, or novel of education, as it applies to Sherston, rather than to the subgenre *Künstlerroman*, or novel dealing with the education and formation of an artist, as it would apply to Sassoon.[28] In Sassoon's literary calculus, Sherston may be a memoirist, but he is not a poet, thus not fully Sassoon, but, rather, Sassoon's "outdoor self" simplified. Sassoon and Sherston may have shared the same experiences, but Sassoon has the advantage of determining which experiences Sherston narrates. Thus, as Sherston announces early in *Fox-Hunting Man*, "The continuity of these memoirs [*Fox-Hunting Man*], is to depend solely on my experiences as a sportsman"; and he adds, later, regarding the relevance of his experiences of the wider world, "Any other interests I had are irrelevant to these memoirs and were in any case subsidiary to my ambition as a sportsman."[29]

Fox-Hunting Man, indeed, traces Sherston's initiation, growth, and acquisition of bona fides not only as a foxhunting man but, more broadly, as a *sportsman*. Part 1 focuses on Sherston as budding "equestrian" and as novice foxhunter, and, later and to a lesser extent, as a competitor in point-to-point races and steeplechases and as a stag hunter. Part 2 finds him returning from boarding school, his first time away from home, with "my cricket bat . . . under my arm." Just as part 1 revolves around Sherston's first hunt, part 2 revolves around the Flower Show Match, "the [cricket] match of the year, and to play in it for the first time of my life was an outstanding event." When he returns home from Cambridge five years later, in part 3, "I had 'taken up golf' and most of my time and energy had evaporated on the links." He reveals only later, in *Infantry Man*, his one lacuna as a sportsman: "I had never shot at a bird or an animal in my life, though I'd often felt that my position as a sportsman would be stronger if I were 'a good man with a gun.'" He acquires that skill, of course, at the front.[30]

Sherston melds his identities as sportsman and military man in *Fox-Hunting Man* and *Infantry Officer* through both chronology and symmetrical structure: each work comprises ten parts, and each plays its first eight parts against its last two. Sherston spends four-fifths of *Fox-Hunting Man* recalling his youthful progress as sportsman in the environs of Butley, the country village where his Aunt Evelyn raised him, and the final fifth recalling his immersion into the military from recruitment in Canterbury, to officer training near Liverpool, to embarkation to France and debarkation on the Western Front. Similarly, he spends four-fifths of *Infantry Officer* recounting his military experiences up to and including being seriously wounded, and the final fifth recounting his convalescence and initiation into very public antiwar activism. The longer first part of each work, as we shall see, stands both in opposition and apposition to the latter part, creating an overall thematic ambiguity whose origin lies in the trope of the hunt and whose end lies in Sherston's transition, to quote Fussell, from "fox-hunting to Boche-hunting."[31]

Almost like getting hunting clothes

Fox-Hunting Man first appeared anonymously (not under the name of a famous Great War poet) and, of course, autonomously (as the first volume, it was not yet part of a Great War trilogy). The reading public in 1928 (like many readers in sporting circles today) received and celebrated *Fox-Hunting Man* as one of many gently ironic memoirs of foxhunting, and paeans to it, published in England and America between the wars.[32] Though Sassoon proffered it as such, he also intended, as became clear two years later, to pair attributes of hunting in England with those of soldiering in France:

Throughout the history of English literature, writers and their works treated rural life, including hunting, as embodiment of timeless traditional values. Sassoon speaks in *Siegfried's Journey* of "my intolerance of the unusual and my instinctive preference for the traditional," and Sherston consistently approves rural tradition, especially hunting, and censures modernity, especially when it disturbs hunting. Likewise, Sherston may lampoon his rural Butley neighbors, but he also defends them in the end: Captain Huxtable, for example, is "an epitome of all that was most pleasant and homely in the countrified life for which I was proposing to risk my own."[33]

Writers also used rural landscape, not surprisingly, as synecdoche not only for the beauty of both uncultivated nature and cultivated land, but also for rural life and its virtues, as well as its limitations. Sherston recalls with nostalgia, "I cannot think of [my own countryside] without a sense of heartache, as if it contained something which I have never quite been able to discover," while Sassoon, without contradiction, attributes Sherston's provincialism before he went to war, and "across the Channel" for the first time, to his rural upbringing, including, as a foxhunter, his intimacy with its countryside.[34]

Foxhunts traversed private farmland, and writers who idealized rural life and land also idealized the farmer. Ever jejune, the young Sherston sniffs, "The country was there to be ridden over. That was all." After the master of his hunt declares "that the best friend of the fox-hunter was the farmer," however, even Sherston comes to see, "These people were the pillars of the Hunt—the landowners and the farmers. The remainder were merely subscribers." Eventually, Sherston converts prewar respect for the farmer into wartime hagiography: "[Corporal Griffiths] was only a stolid young farmer from Montgomeryshire; only; but such men, I think were England, in those dreadful years of war."[35]

The symbolic value assigned to the land and its stewards led inevitably to the trope of "waste land," fully realized in Eliot's *The Waste Land* and subsequently echoed by other war writers. Sassoon first invokes the trope with reference to training camps in England—"most of them were constructed on waste land; and to waste land they have relapsed"—and later applies it to the trenches in France—"we seemed to be walking in a waste land where dead men had been left out in the rain after being killed for no apparent purpose."[36] Like Eliot, however, Sherston also holds out for farming as redemption of the land, and, by metaphoric extension, for agriculture as redemption of culture.

As a defining pastime of British aristocracy and landed gentry, hunting incorporated elements of class structure, including defined internal hierarchy, rigidly enforced etiquette, and studiously practiced jargon. Hunting also served farmers as a means for removing nuisance predators, however, so it crossed social lines defined, respectively, by leisure and husbandry. A hunt, depending on its constituency, might reenact social inequality—the master of the Ringwell Hunt "was a rich man whose . . . only qualification was his wealth"—or might mitigate it: "And since the field [of the Coshford Vale Stag Hunt] was mainly composed of farmers, there was nothing smart or snobbish about the proceedings."[37]

Whatever the case, foxhunting as a conveyer of skill and vessel of tradition conferred status on the foxhunter, particularly while a hunter served in the officer ranks of the military, mainly drawn from the upper classes.[38] Sherston discovers "that it was a distinct asset, when in close contact with officers of the Regular Army, to be able to converse convincingly about hunting. It gave one an almost unfair advantage in some ways."[39] Moreover, "I was readily given leave off Saturday morning duties [to ride to hounds], since an officer who wanted to go out hunting was rightly regarded as an upholder of pre-war regimental traditions."[40]

A hunter's attire, like an officer's uniform, signaled to outsiders the wearer's standing in an established order, while the quality of tailoring signaled to insiders his or her means and taste. Sherston, who as a youth "enjoyed dressing up as a sportsman," and who supposed on buying his first hunt clothes, "In outward appearance, at least, I was now a very presentable foxhunter," is soon disabused: seasoned hunters cared more about his inward mettle. The lesson took: though finding later that "ordering my uniform from Craven & Sons was quite enjoyable—almost like getting hunting clothes"—Sherston also denounces, if only in retrospect, officers who cared more about his appearance than his character.[41]

In addition to attire, tack and other accoutrements also conveyed status in the hunt and military—status accruing from aesthetic quality, practical utility, or both, as Sherston's repeated references to boots suggest. Young Sherston's "new [hunting] boots," like the "well-cut riding boots" of young cavalrymen, carry explicit symbolic value. Officer Sherston's costly "brown Craxwell field-boots" and "pair of greased marching boots whose supple strength had never failed to keep the water out" have explicit practical value, but also symbolize distinctions in class and rank between well-shod officers and troops wearing "bad boots."[42]

Symbolic value notwithstanding, neither clothes nor fine boots can substitute for the essential traits of pluck and prowess demanded by the shared hazards of the hunt and the front—two traits that define Sherston's identity . . . and his identity crisis. Robert Graves captured the essence of those hazards when he wrote that "mountaineering can be much safer than foxhunting. Hunting implies uncontrollable factors, such as hidden wire, holes in which a horse may stumble, caprice or vice in the horse. Climbers trust entirely to their own feet, legs, hands, shoulders, sense of balance, judgment of distance."[43]

With those basic points, Sherston provides the background for his creator's (and thus his own) character. Wohl kens it nicely: "Sassoon enjoyed fighting, as long as he could go about it his own way, as a sportsman."[44]

The hunting organism

Sherston's character, certainly as he represents it, owes to his fastidiously cultivated identity as foxhunter and sportsman. "Had my Aunt Evelyn employed an unpretentious groom-gardener" as her coachman, rather than "jaunty young Dixon," Sherston begins his narrative, "I should never have earned the right to call myself a fox-hunting man." Without the mentorship of Dixon, for whom "the Kennels were the centre of the local universe," Sherston would not have become "a sportsman" or have embarked on "my career as a fox-hunting man" who measured the "progress" of his early "terrestrial experience" by the fact that "I had averaged five days a fortnight with the hounds," and who depicted himself in his diary "as I wanted myself to be—a hard-bitten hunting man." Far more than a sport or lifestyle, in short, hunting for Sherston is a metaphoric life form that a horseman joins but that, in the end, owns the horseman: "The hunting organism [was] the only one worth belonging to."[45]

In addition to such fealty, the hunting organism also demands adroitness. Sherston fears, particularly as an acolyte, that "my performance had consisted not so much in riding to hounds as in acting as a hindrance to [his horse's] freedom of movement," a fear that he assuages by slighting inept or timid riders: "It was my wilful habit in those days to regard everyone who preferred going through a gate to floundering over a fence as unworthy of the name of sportsman." He also assuages his fear, not surprisingly, by exalting adept riders—especially Denis Milden, a boyhood crush and later master of the hunt and friend: "His eyes were on the hounds, and he went over the country, as we used to say, 'as if it wasn't there.'" When in the military, Sherston apportions skill with equal punctilio: senior officers, variously commissioned, generally lack competence, and troopers, insufficiently trained, generally lack efficiency, but subalterns, like himself, compensate by bringing skill and mettle acquired elsewhere—namely, in sport.[46]

The hunting organism, finally, also demands ethics, or, in Sherston's worldview, sportsmanship. Though critical to ensuring the cohesiveness and integrity of sporting and military entities alike, Sherston intimates, ethical standards gain more purchase if organically evolved and willingly adopted, as in hunting, rather than mechanically developed and willfully imposed, as in the military. In either case, they depend on tradition for effectiveness. The worst offenders in *Fox-Hunting Man*, for example, are farmers who "made no secret of shooting" foxes, cited by Sherston as prima facie evidence that Aunt Evelyn lived in a "thoroughly unsporting neighborhood." Sherston probes but never resolves, however, if that is a distinction without a difference. The fox, that is, dies whether shot or chased, and from a much bloodier death in the latter, a lesson in the ethics (and aesthetics) of violence that Sherston, who once "felt an unconfessed sympathy" for his quarry, will carry from the hunt to the front.[47]

Whatever the case, the hunting organism demands, above all, a boldness and courage that Sherston tracks directly from the hunt to battle to his public antiwar statement—a linkage that Sassoon made explicit when he wrote in *Siegfried's Journey* that his own statement (the real-life basis for Sherston's) "appeared to be a moral equivalent of 'going over the top,'" both actions "requiring moral courage." Sherston recalls that "courage remained a virtue" to him when he enlisted, but, as he later discovered, the "exploitation of courage, if I may be allowed to say a thing so obvious, was the essential tragedy of the War." Though courage can be exploited, even betrayed, that is, Sherston is arguing that the hunter, soldier, and resister alike must possess and display true courage—not thinking, for example, of "trench warfare as an adventure" or mistaking "bravado [for] bravery"—particularly in the face of "uncontrollable factors." Boldness in doing so seems to require, above all else, willingness to accept the irrevocability of one's actions, as cavalrymen long had insisted.[48]

Sherston introduces this theme of irrevocable action by recalling his friend Stephen's starting a point-to-point race while simultaneously wishing that it "was all over, or that something would happen to prevent it taking place at all"—prompting Sherston's mindful observation "that without such feelings heroism could not exist." Once on course, in any case, "Stephen was then turned adrift with all his troubles in front of him. No one could help him any more," a helplessness that Stephen shares with his horse's owner, Colonel Hesmon, whose bearing Sherston now sees as a "premonition" of wartime departures: "Elderly people used to look like that during the War, when they had said good-bye to someone and the train had left them alone on the station platform." Similarly, and more important, Sherston's facing a big hedge on an early hunt with the feeling "that I was 'in for it'" foreshadows his fatalism when writing his antiwar statement—"my heart was beating violently. I knew that I couldn't turn back now"—and when reporting by train to his commander after sending it—"I completely lost my nerve. But the express train was carrying me along; I couldn't stop it, any more than I could cancel my statement."[49]

Horses, humans, and holes

Hunting, in its essentials, comprises mounted humans deploying and following a canine pack chasing vulpine prey across a rural landscape. Putting aside hounds and foxes, Sherston treats each of the other components literally in *Fox-Hunting Man*, and also employs each of them, in that volume and, especially, in *Infantry Officer*, as symbolic of aspects of the Great War and his roles in it, though he tends to use these symbols loosely rather than build a cohesive analogy between the particulars of a hunt and a battle.[50] Whether as mount and partner in the hunt, or as resource to cavalry, artillery, and infantry, horses provided the motive power. Whether as members of a community or as individuals in an organization—adepts in a hunting organism or cogs in a military machine—humans served as principals. And whether as rolling Elysian farmland or cratered hellish waste land, terrain created both enabling and constraining conditions.

Not surprisingly, *Fox-Hunting Man* devotes considerable attention to horses and horsemanship. Sherston invokes and talks about each of his serial and increasingly capable horses: the pony Rob Roy, the Welsh cob Sheila, and his field hunters Harkaway, Cockbird, and Sunny Jim.[51] He also makes much of his enlistment in the army and Cockbird's concurrent requisition: Sherston was not convinced "that I ought to become an officer myself," but, as a responsible horseman, he transfers Cockbird "to the squadron commander" (to ensure lighter work), so Cockbird "had in a manner of speaking, accepted a commission." Though government policy would allow Sherston to buy Cockbird back after the war, "I knew that I had lost him." Cockbird's requisition and Sherston's sacrifice of him symbolically represent not only the latter's transition in "career" from horseman to soldier—"it was a step nearer to bleak realization of what I was in for"—but also the fate of untold numbers of "the horsehood of England" acquired and wasted during the war.[52]

That wastage informs three equine themes in *Infantry Officer*. One, horses served as cavalry chargers, to be sure, but primarily in artillery and transport roles where they suffered mass carnage through shelling, as well as individual terror: "Now came an interval of silence [during a shelling] in which I heard a horse neigh, shrill and scared and lonely." Two, as a result, equine corpses, conspicuous among "the horrors of war," littered the landscape by the thousands: Sherston speaks of dead soldiers and "dead horses . . . nearly three days old," just as Sassoon speaks in a poem, "The Road," of equines lying where they died, "big bellied horses with stiff legs." And three, equine recruits, unlike human, had neither voice nor choice, a common theme, as we saw previously, in Great War representations. As Graves wrote, "I was shocked by the dead horses and mules; human corpses were all very well, but it seemed wrong for animals to be dragged into the war like this." While the human corpses, especially those of "common soldiers," equaled equine corpses in number and horror, and while conscripted men, like requisitioned horses, were victims, the conscripts had at least some voice and an illusion of choice.[53]

As a fictional autobiographical memoir recalling the life of an individual human among humans—alive and dead—Sherston's trilogy portrays a range from a handful of main characters, to many incidental characters, and also, particularly in *Fox-Hunting Man*, several "types" familiar in the literatures of country life and of war.[54] Sherston as memoirist tends to contrast hunting types sharply: he can be treacly about sportsmen, such as the exemplary Dixon, and acrid about country louts, such as Bill Jaggett, "to my mind, one of the horrors of the Hunt."[55] He tends to contrast enlisted types more gently, however, by their ways of coping with shared horrors: Rees was "garrulous and excitable [and] uncouth, and he made no pretensions to being 'a gentleman,' [while] Shirley, true to the traditions of his class, simulated nonchalance, discussing with Leake (also an Oxford man) the comparative merits of Madgalen and Christ Church." The hunting and military entities, in any case, have the final word: the hunt tolerated restiveness despite protocols when the "ungovernable" subscribers were "brilliant riders," and the war crossed class boundaries where death was concerned: "Both Shirley and Rees were killed before the autumn."[56]

Setting Sherston and those characters in two homosocial environments, the trilogy, like many Great War memoirs, wears homoeroticism on its sleeve, though it tracks its protagonist's affective (and obliquely sexual) life to foxhunting rather than to public school or university. It incorporates, for example, and as Fussell notes, common set pieces like soldiers bathing and common types like the blond, handsome, and doomed young warrior. Sherston's specific objects of desire—Denis Milden in *Fox-Hunting Man* and Dick Tiltwood in *Infantry Officer*—play transformative roles in his progress. Though Dixon introduces Sherston to the sporting life, Milden seduces him fully into it: "I was being magnetized to a distant meet of the hounds, not so much through my sporting instinct as by the appeal which Denis Milden had made to my imagination." The appeal grows even stronger when Milden returns as master of the hunt. Though Milden's wartime counterpart, Dick Tiltwood, is a less fully developed character, his death in action prompts Sherston, then older and more passionate, to launch the "suicidal" raid that results in his wounding, convalescence, and, ultimately, antiwar statement.[57]

Fox-Hunting Man also embeds a related theme of "manliness" even more pertinent than homoeroticism. Vexed by "an 'unmanly' element in my nature," young Sherston, his older self recalls, immediately recognized that Milden "had the voice of a boy, but his manner was severely grown-up"—meaning, of course, the manner of a "man." Sherston, then, is mortified when he commits an "indiscretion" on seeing his first fox, and "I knew only too well what a molly-coddle I had made myself in the estimation of the proper little sportsman [Milden] on whom I had hoped to model myself." Some years later, though chiding an older hunter who "had conventional eighteenth-century ideas about what constituted masculine gallantry" that made him "a complete anachronism," Sherston describes three young hunting brothers as "reckless, insolent, unprincipled, and aggressively competitive, but [also] . . . desperately fine specimens of a genuine English traditional type which has become innocuous since the abolition of duelling." Overt in *Fox-Hunting Man*, the theme becomes conspicuous by its absence from *Infantry Officer*: ideals of "manliness" that prevailed in the country are moot at the front.[58]

A More Than Half-Made Soldier

Fox-Hunting Man and *Infantry Officer* obviously place foxhunting and warfare in counterpoint. They contrast the freedom both realized in and symbolized by hunting—the experience of fast and exhilarating movement on one's horse over bright and broad fields dotted with woods, hedges, and fences—with the confinement equally realized in and symbolized by trench warfare—the experience of cramped and terrified movement in dark and narrow trenches or on one's belly through corpse-strewn mud and "derelict wire." The two volumes and their successor, *Sherston's Progress*, moreover, juxtapose freedom and confinement literally in the physical freedom of Sherston's "leaves" from action and the mental freedom of his vocal pacifism, as opposed to his confinement in a hospital for shell-shock victims for having spoken his mind.

At the same time, however, the first two volumes of the Sherston trilogy explicitly, and the third volume implicitly, also treat hunting and warfare as counterparts. Sassoon, through Sherston, for example, refers to hunting as "inhumane" and war as "inhuman"; he intimates that both enact primitive ritual killing despite their rational and utilitarian justifications (though he insists less on this point than Hemingway does on bullfighting and warfare); and he indicates that both entail attrition of men and horses (though in hunting attrition is incidental and often temporary, but in warfare intrinsic and often permanent).[59] As we have seen, Sassoon also develops many parallels between the roles of men and horses in the two endeavors. Bifocal in perspective, in short, the trilogy is also bivalent in conviction: it presents hunting and warfare, at once, as counterpoints and counterparts, with irony as the uneasy mediator.

Sherston's irony runs the gamut. He pokes light fun at a Butley "type," Colonel Hesmon, who "patterned himself [on notions that] were part regimental and part sporting," but who had "never seen any active service" and lacked both skill and boldness in the field; he engages in gentle self-irony, "astonished" to learn that "those damned socialists who want to stop us hunting . . . opposed conscription as violently as many fox-hunting men supported the convention of soldiering"; but he casts a gimlet eye on an imagined hospital visitor, *Hunting Friend (a few years above Military Service Age)*: 'Jokes about the Germans, as if throwing bombs at them was a tolerable substitute for fox-hunting.'" Sherston's irony becomes harsher, though, when he parallels his excitement at first hearing the thrilling sound signifying "that one of the whips has viewed the fox quitting the covert" with the "series of 'view-halloas'" that he almost unconsciously emits when assaulting a German trench, "a gesture which ought to win the approval of people who still regard war as a form of outdoor sport."[60]

Sherston's cognitive dissonance in seeing the hunt and the war as both alike and unalike informs his recollection of hunting in Butley while on leave, "listening to Colonel Hesmon while the hounds are being blown out of a big wood—hearing how well young Winchell has done with his Brigade (without wondering how many of them have been 'blown out' of their trenches)." It likewise informs his first comparing himself as a sportsman "ambitious of winning races" with himself as an officer who "wanted to make the World War serve a similar purpose" and "get a Military Cross," and then contrasting both those selves with the resister who threw his MC ribbon into the Mersey: "One of my point-to-point cups would have served my purpose more satisfyingly, and they'd meant much the same to me as my Military Cross." He never clarifies the precise comparative value, if any, between the cup and the medal, or, for that matter, his antiwar gesture of casting away the decoration.[61]

Nothing but hunting, racing, and polo

In addition to dissonance, Sherston experiences leaden oppression from the losses and "imprisonment" that pervade his wartime narrative. When he loses Cockbird, to take an early example, he loses "my only tangible link with the peaceful past"; and when he learns Kitchener's prediction of a prolonged war, at about the same time, he anticipates with dread "three years of imprisonment." His tropes of loss evoke amputation and atrophy, both in fact endemic in the Great War, and those of imprisonment, especially "the crouching imprisonment of trench warfare," both physical and mental confinement. Despite Sherston's paradoxical and somewhat hollow claim of going on dangerous patrols as "the only escape into freedom which I could contrive," and of finding that "I could at least escape from the War by being in it," Sherston actually treats both loss and imprisonment with the same salve: country pursuits.[62]

That salve comes in three distinct but related forms. The first, reminiscing, is purely mental, internal, and private and involves cricket and hunting equally. Evoking cricket as "an epitome of all that was peaceful in my past" and as synecdoche for "an England where there was no war on," Sherston thinks "about driving home from cricket matches before the War, wondering whether I'd ever go back to that sort of thing again," and reminisces about "April evenings in England and the Butley cricket field." Similarly, "Meditating about England . . . I thought of the huntsman walking out in his long white coat with the hounds" (as opposed to himself standing "in the front-line with soaked feet"); and "The view across the Weald at sunset," from a country home where he convalesces, "had revived my memories of 'the good old days when I hunted with the Ringwell.'" Since Sherston ascribes the same reminiscing to others, picturing "our second-in-command," for example, "consoling himself with reminiscences of cricket and hunting," it represents a common, but still private, effort at coping, one also evoked in Sassoon's war poems.[63]

When reminiscing becomes conversing, however, it becomes a shared interpersonal effort, and, in *Infantry Officer*, that effort revolves almost entirely around hunting. The sharing can be epistolary, as in Stephen's letters to Sherston, or interlocutory, as in Sherston's many verbal exchanges with various individuals. The latter might be old hunting friends with whom one can share reminiscences and restore one's "mental equilibrium." They might be casual wartime acquaintances with whom one is just passing the time "jawing." Or they might be very close wartime friends with whom one is intellectually and emotionally intimate, such as Dick Tiltwood or David Cromlech. Given the numbers of sportsmen in an officer cadre drawn mainly from the upper classes, such conversations also could be more communal (even if among a coterie): "There were some fine riders in the regiment then; they talked and thought about nothing but hunting, racing, and polo." Whether monologic, dialogic, or polylogic, in any event, reminiscing and conversing are *about* something, not the thing itself.[64]

Sherston critically adds *doing*, then, to the mix, in effect realizing his wish to "wake up and find myself living [in a Trollope novel] and hunting three days a week." When in France, he "borrowed the little black mare no one could ride and cantered about the open country by myself . . . two or three afternoons a week," sometimes engaging in "an imitation hunt" with "imaginary" hounds and foxes. (Graves likewise reported in *Goodbye to All That* "being free to ride all my three chargers over the countryside" when in France.) More important, when on leave, Sherston continues to hunt both as pastime and as symbol of his core identity. In the first significant wartime instance of hunting, Sherston, having been wounded, returns to Butley and "some mornings with the hounds" of the Ringwell Hunt until, inevitably, the final day's meet ends, and "I went one way and the hounds went another. . . . They disappeared in the drizzling dusk." Back at camp before shipping out, "The contrast between Clitherland Camp and the Cheshire Saturday country was like the difference between War and Peace."[65]

The more significant instance, narrated in *Sherston's Progress*, finds Sherston on leave in Limerick, Ireland, after his rebellion and subsequent return to service. Reminded that when "galloping and jumping on a good horse everything else was forgotten—for forty-five minutes of the best, anyhow," he joins the Limerick Hounds, where "everyone rode as if there wasn't a worry in the world except hounds worrying foxes." That "escape" too, however, must end—"how kind they were, those friendly fox-hunters, and how I hated leaving them." The final scenes of the trilogy find Sherston not hunting, but riding "on the company charger . . . 'a solitary horseman' [who felt in riding] a sort of personal manifesto of being intensely alive. . . . That was how active service [also] used to hoodwink us. Wonderful moments in the War, we called them, and told people at home that after all we wouldn't have missed it for worlds." This war of deceit and betrayal, he realizes at the end, compromises everything, including and especially nostalgia.[66]

"Ware wire!"

As noted earlier, Sassoon juxtaposes the firm ground and open fields of the hunt to the waste land of the front—an "enemy world [with] no relation to the landscape of life," a place blanketed in mud and gore, dense with impregnable tangles of wire, and pocked with craters some "fifty yards in diameter and about fifty feet deep. . . . Their sides were steep and composed of thin soft soil; there was water at the bottom of them," and men often drowned in it. Employing the archetype of *above ground / below ground*, with its ancient associations of world and underworld, life and death, he is not out to distinguish foxhunters from their vulpine prey, but rather to distinguish them from soldiers fighting from trenches, burrowing through tunnels, crawling in craters, and huddling in dugouts. "And when all is said and done," Sherston concludes, "the War was mainly a matter of holes and ditches." It also was a matter of something else that serves Sassoon as a pivotal metaphor: wire.[67]

Sassoon employs the metaphor nearly a dozen times in *Fox-Hunting Man*—about evenly divided between the hunt and the war. Farmers who employ "the most dangerous enemy of the hunting-man . . . barbed wire"—the reference to "enemy" surely not accidental—are "blighters" thwarting the hunt, such as "a double-distilled blighter who's wired up all his fences"; and riders who use wire as an excuse are "blighters" disrupting the hunt, such as a member of Sherston's hunt who avoids taking a small fence "with a shout of 'Ware wire!' . . . the wire having been an improvisation of [his] over-prudent mind." Even more important, although

Sherston envisions "an Elysium of green fields and jumpable hedges . . . and in those days there was very little wire in the fences," he takes a fall "at a lush-looking obstacle [because] I failed to observe that there was a strand of wire in it." The cause of the fall—failing to see hidden wire— obviously symbolizes the distance between Sherston's ideal vision of country life and the hidden realities that, as a consequence, he fails to see. And the resulting fall and broken arm, delaying but not preventing his enlistment, foreshadow his recognition of that distance as it applies first to Butley and later to the war—a fall from grace, or from innocence into experience.[68]

At the front, Sherston joins his comrades in the dangerous and often fatal duty of laying and repairing the wire that guards their trenches. That duty resulted in the killing of Dick Tiltwood, "hit in the throat by a rifle bullet while out with the wiring-party. . . . I knew Death then."[69] Sherston resigns his assignment as transport officer (and thus an assignment involving horses) soon after—late in April, "during the last thirty days of Lent," he points out, and transforms into "Mad Jack," the homicidal and suicidal avenger: "I went up to the trenches with the intention of trying to kill someone. It was my idea of getting a bit of my own back," and being "ready for any suicidal exploit."[70] The final scene of *Fox-Hunting Man* finds Sherston emerging from his dugout into the trench. "I stared at the tangles of wire and the leaning posts, and there seemed no sort of comfort left in life." Neither nostalgia, in the form of a reverie of Butley, nor belief offers that comfort: "I remembered that it was Easter Sunday. Standing in that dismal ditch, I could find no consolation in the thought that Christ was risen." Sherston, who had risen, injured, from his literal fall, has yet to rise, redeemed, from his symbolic fall.[71]

Infantry Officer then expands the metaphor of laying the wire that guards one's own trenches to the even more hazardous task of cutting the wire that guards the enemy's trenches—the primary objective of wire-cutting patrols and the necessary but secondary objective of raiding parties. The metaphor first serves social critique, enabling Sassoon to draw a distinction between officers and enlisted men: after "cursing the bad wire-cutters which had been served out for [a] raid," Sherston, while on leave, "invaded the Army and Navy Stores and procured two pairs of wire-cutters with rubber-covered handles . . . my private contribution to the Great Offensive . . . very civilized [wire-cutters], which looked almost too good for the Front Line." A distinction in financial means as well as in aesthetics, it separates men not only by their rank, but also by their class—in the end, the same thing—a distinction perhaps underscored when David Cromlech "announced that he'd been doing a bit of wire-pulling on my behalf" following Sherston's antiwar statement. The quality wire cutters that Sherston breezily calls a "contribution"—in effect, a gesture of noblesse oblige—an enlisted man might call an unobtainable life-saving tool.[72]

More important than enabling social critique, however, the metaphor of wire advances the theme of moral awakening and progress toward redemption. Sherston's progress toward redemption first glimmers when he exchanges vengeance for beau geste: "I had made up my mind to have another cut at the wire, which I now regarded with personal enmity," and to do so, despite the added danger, "by daylight because commonsense warned me that the lives of several hundred soldiers might depend on it being done properly." His progress shines even brighter during his convalescence. "The bitter reality [of the war] returned to me as I squeezed myself through the hospital's barbed wire fence" to walk peacefully in the woods. "I was losing my belief in the War, and I longed for mental acquiescence." Still longing for emotional and spiritual escape, put differently, Sherston also must find *intellectual* escape: the curse neither of the sportsman nor of the soldier, but of the ironist.[73]

Sherston the older memoirist is that ironist, but the journey through his long narrative seems to lead him, in the late scenes of *Sherston's Progress*, *not* to deploy that rhetorical weapon. He conjures himself after returning to the front, trying to understand, as cited earlier, "whether there was any meaning in this human tragedy. . . . I took off all my equipment, strolled along to the nearest sentry, borrowed his bayonet, and told him that I was going out to have a look at the wire." Instead, he crosses no-man's-land, "crept through a few strands of wire," and came upon a German machine-gun team: "I was at last more or less in contact with the enemies of England. . . . And what I saw was four harmless young Germans who were staring up at a distant aeroplane."[74] The wire that caused the foxhunter Sherston to fall from grace also prompted the officer Sherston, on Easter Sunday, to recall but not regain his lost paradise. The wire that robbed Sherston of love in the body of Tiltwood and left him as the vengeful Mad Jack, also allowed his "leaves" and a peaceful vision of shared humanity with an upward-looking enemy.

The hunting breed of man

If wire is the metaphoric strand that binds Sherston's trilogy, the idea of hunting and warfare as parallel endeavors is the extended trope that provides its foundation—a parallel that Sassoon, clearly conflicted, both subverts with irony and reinforces with nostalgia. Far from inventing the trope, Sassoon is drawing on a well-established tradition in British military writing—one that his reader, he could assume, would know to one degree or another. For obvious reasons, writers employing this parallel applied it mainly—though not exclusively—to the kinship of hunting and cavalry, using the parallel literally to argue for hunting as ideal preparation for the cavalry officer—what Major-General M. F. Rimington, in *Our Cavalry* (1912), called "that best of schools the hunting-field."[75] That idea, not surprisingly, emerged and evolved over the eighteenth and nineteenth centuries, the great age of British cavalry. It also extended well into the twentieth century, sometimes literally, more often as a potent trope.

Initially, in the eighteenth century and through the Napoleonic period, when cavalry maneuvers depended on precise movements of large bodies of horsemen in closed ranks, writers such as William Tyndale, as we saw, emphasized the limitations of the national style of cross-country equitation derived from foxhunting, a problem, in Frederick Maude's opinion, that obtained far into the nineteenth century.[76] Over that century, however, cavalry tactics changed, writers recognized similarities between the new tactics and foxhunting, and they began to argue in favor of hunting as fit preparation for cavalry; by late century, they had reached consensus on the principle of combining *manège* and cross-country riding. Advocacy for foxhunting as preparation for cavalry had two rationales. At one level, writers emphasized the skills that hunting conferred on the cavalryman—the horsemanship skills of assessing, understanding, handling, and riding horses competently and confidently, and the field skills of "reading" terrain instinctively and accurately and "riding" terrain (as Sherston says of Denis Milden) "as if it wasn't there." At a second level, writers emphasized the character and mettle that hunting fostered and that cavalrymen, particularly officers, needed to possess and display.

With respect to skills, Lewis Edward Nolan, an early advocate of foxhunting and *manège* training as complementary preparations for cavalry, wrote in *The Training of Cavalry Remount Horses* (1852), "Few of our

[British] cavalry officers would be stopped by a fence; but for this they are not indebted to what they learn in the riding school, but to their being accustomed to ride across country. All foreign cavalry [by contrast] practise at the *leaping bar*, yet their officers, when they meet with a wall or a gate, are *pounded*." As M. Horace Hayes would concur a half-century later, in England "riding means riding to hounds, which has not much in common with school performances." Between the Boer and Great Wars, Field Marshal Earl Roberts, introducing a work by Erskine Childers, proposed, "The officers should possess all the qualities of good sportsmen. They should be fine riders, careful horse-masters, have a keen eye for country, and be thoroughly well educated." And two years later, Rimington added that "riding to hounds over difficult country [gives] an officer an eye for country, and a decision in crossing it, unobtainable in any other fashion."[77]

Skills, however, were only the half of it. In the minds of cavalry authors, hunting also imparted mettle—or, as Sherston has it, "manliness"— revealed in "boldness."[78] Since these authors were writing explicitly to and for officers, moreover, they used the hunt and its imparted mettle to define the "type" who could meet the demands of cavalry leadership and to identify those who would qualify for it. Continental adversaries may have taken exception, but British cavalry officers from the Napoleonic Wars up to and after the Great War, though differing on the hunt as imparting the right riding skills, uniformly saw it as building character. Rimington, the most zealous prewar British advocate, found his ideal type in "a man who combines an addiction to, and some knowledge of, field sports, involving horses, with sufficient intelligence to pass into Sandhurst. . . . We particularly want the hunting breed of man, because he goes into danger for the love of it"—the same ideal invoked by postwar American advocate, Malcolm Wheeler-Nicholson, in *Modern Cavalry* (1922): "The man who rides into danger for the love of it, the man who keenly enjoys cross-country going and polo, contains in his disposition the germs of success as a cavalry officer."[79]

Rimington and Wheeler-Nicholson each defended that position at length and with zeal. Rimington contends, "The addiction to manly, and especially to rough and dangerous, field sports must be regarded as an immense asset towards efficiency for war. Time spent in the chase, 'the image of war,' must not be regarded as so many hours less given to his employer by the cavalry officer"; he prescribes "that the young officer should be for choice country bred, fond of sport, a 'trier,' and must have some private income"; and he affirms, "It is a duty to his country for a cavalry officer in peace-time to take such exercise in the available sports of hunting, pig-sticking, polo, big-game shooting, and other [equestrian] exercises. . . . The cavalry officer . . . who seldom gets on a horse in peace-time, will not suddenly change his nature in war." Wheeler-Nicholson, similarly, advises, "Effort should be made to secure as reserve officers . . . young men whose interests are allied with cavalry interests—horse breeders, polo players, gentlemen jockeys, and horse enthusiasts generally"; and he argues that requisite cavalry proficiencies include "love of the horse" and skill in "the mounted sports," especially polo and fox hunting.[80]

British nationalism inflected this theme from the start. Nolan contended not only that "the Englishman beats the world in a ride after the hounds and a run across country [in] this manly sport—the best of all to form bold riders," but also that English women, "would beat, on the field, all [the Continent's] mathematical riding-masters, and take gates, fences, and ditches, from which foreign officers of hussars or their dragoon roughriders would turn aside in dismay to look out for a break or gap." Sir Evelyn Wood, likewise, boasted in *Achievements of Cavalry* (1897), "We have one incalculable advantage which no other nation possesses, i.e. that our officers are able to hunt." While George T. Denison declared in *Modern Cavalry* (1868) that English hunting was more likely than American racing "to produce the best horseman," Theodore Roosevelt, in "Riding to Hounds on Long Island" (1886), transposed the sentiment from British nationalism to American sectionalism: "If in 1860 riding to hounds had been at the North, as it was at the South, a national pastime, it would not have taken us until well on towards the middle of the war before we were able to develop a cavalry capable of withstanding the shock of the Southern horsemen."[81]

Since military authors believed themselves defending essentially immutable values, their advocacy for the virtues of foxhunting neither wavered in commitment nor altered in substance over time. It arguably reached its greatest height of enthusiasm and width of application in E. A. H. Alderson's *Pink and Scarlet, or Hunting as a School for Soldiering* (1900), whose preface carries a notation identifying the book's composition aboard "S. S. 'Malta,' *En route to South Africa, October 1899.*" Alderson's book is especially relevant to the Sherston trilogy because it treats hunting, as its title indicates, as preparation not just for cavalry, but rather for "*soldiering*" in all arms. If "the hunting man is already a more than half-made soldier," as Alderson contends, then addressing the other half with the "things that hunting cannot help teaching, and the many, many things it *may be made to teach if taken in the right way*," should complete the job of making him whole. Though *Sherston* describes that process and progress with the same irony and humor that Alderson employs, *Sassoon*, in the end, may not share fully either the irony or the humor.[82]

Richard Greville Verney, Lord Willoughby de Broke, began "The Future of Fox-Hunting," the opening chapter of his "reflections," *Hunting the Fox* (1920): "When we declared war upon Germany in 1914, many people thought that Fox-hunting in the British Isles was doomed," but, instead, mobilization resulted in "recognition of Fox-hunting as a first-class national asset. . . . The Expeditionary Force could not have left England unless the nation could have drawn upon studs of well-bred hunters to bring the Peace establishment of Army horses up to war strength." Verney then adds the real point of his book: "But quite apart from . . . national utility, Fox-hunting will surely survive from its own innate qualities. The manner in which it has lived through all the obstacles of war time is a sufficient testimony to its vitality." Less confident of that vitality a decade later, Ladies Diana Shedden and Apsley, in "*To Whom the Goddess . . .*" (1932), offered a litany of reasons for why "we now face bad times," naming "the elimination of the great landowning and yeoman families . . . the spread of bungaloid towns into the country [and] the agitation of the Anti-Blood Sports people" as among "the many difficulties which confront those who are trying so well to uphold the great traditions of the Past." To dismiss these writers simply as social anachronisms would be a mistake.[83]

The Great War had rendered horse cavalry on a large scale, for most practical purposes, a military (and also political) anachronism, effectively ending literal appeals to "riding to hounds" as preparation for soldiering skills and battlefield mettle, but not, in the end, foreclosing figurative uses of the parallel. A plethora of books published at precisely this moment, Verney's and Shedden and Apsley's among them, reaffirmed the foxhunt as icon of a disappearing rural British culture and the traditional values instilled in its huntsmen and fields: respect for tradition and for boldness in character. Sassoon's *Memoirs of George Sherston*, finally more elegiac than ironic, respected that culture and both reflected and promoted a profound sense of loss over its passing. Sassoon, in short, revived an old, moribund, and particularly British conflation of field sports and mounted warfare; he buried that conflation under three volumes of irony; but, in the end, he resurrected it. His Sherston trilogy embodies interwar nostalgia for prewar country life and exemplifies the Great War literature, and culture, that both questioned and perpetuated it.

In Memoriam. Raffaele Salvatore. Massachusetts. Pvt, 8 company, 151 depot brig, WWI. Born 1896 died 1967.

Postscript

For six millennia, the horse has played a signal role in human history and, for the past half millennium, a decisive role in military history. In the modern era in Europe, from the Renaissance to the twentieth century, mounted warfare depended upon the formal training of military horses and riders, and, for at least two centuries, refined systems of *manège* dressage and equitation provided the theoretical and practical bases for that training. Military firepower steadily evolved over the long nineteenth century, however, and more powerful and accurate guns and rifles both demanded and enabled new mounted tactics and, it follows, training for them. Cavalry complemented and combined *manège* training of horse and rider with bolder forms of cross-country riding derived largely from foxhunting. In a word, systematic *manège* dressage and equitation had been appropriate and adequate for large cavalries of heavy horse engaged in close battlefield maneuvers, but *manège* training of horse and rider together with cross-country training was required for light-horse and dragoons often serving in small detachments over broad distances in the duties of security and reconnaissance.

Horse cavalry, in short, by this point exclusively light-horse and dragoons, had become a highly mobile arm often acting independently, and the image of "daring" cavalry exploits executed by "dashing" cavalrymen imbued with the "cavalry spirit" took hold in the military and public imaginations. That image altered, but it did not disappear, when dismounted fire action first supplemented, then complemented, and finally displaced mounted shock action as the primary role of cavalry, and the carbine (and in local instances the revolver) did the same regarding the saber. The image, as well as the reality behind it, was informed by foxhunting, particularly in England. As Brereton puts it, "Not only did foxhunting breed a thrusting, hard-riding cavalryman, undaunted by any obstacle, but it also saw the evolution of that ideal type of cavalry horse, the hunter." While the Great War essentially rendered actual horse cavalry an anachronism, as Brereton added, "The 'cavalry lobby' [even in the 1930s] could not, or would not, admit that the armoured fighting vehicle must sooner or later supersede the horse as the cavalryman's mount. . . . Only the horse, they claimed, could foster those qualities of boldness and initiative that were as essential to a cavalryman in the field as to a foxhunter across country."[1]

The future demise of horse cavalry signaled by the Great War, together with the mass mechanized slaughter of the war, issued in postwar nostalgia not only for the prewar image of "smart" cavalry, but also for a prewar vision of a vibrant and idyllic country life exemplified by the foxhunt and eulogized by writers like Siegfried Sassoon. In the same interwar years, cavalry officers, who traditionally had been avid foxhunters and polo players, came to dominate international equestrian competition as well, particularly in show jumping and combined training (dressage testing, roads and cross-country riding and jumping, and stadium jumping). Originally called "*le Militaire*" and now known as "The Three-Day Event," the sport evolved directly from cavalry training of horse and rider, gained stature and status when introduced into the Olympic Games of 1912, and, for many years, showcased cavalry officers and horses.[2] In that milieu and moment, and after four centuries, horsemanship began a rapid transition from a primarily military discipline to the primarily civilian activity that it remains today.

Lieutenant-Colonel Harry D. Chamberlin (1887–1944) illustrates that transition. Considered "the finest horseman in the army . . . the very beau ideal of a cavalryman," by Lucien Truscott, Chamberlin exemplified the interwar "cavalry officers [who] usually formed the hard core of Olympic riding teams." A graduate of the American, French, and Italian Cavalry Schools, Chamberlin rode in the Olympic Games of 1920, 1928, and 1932, as well as in countless national and international competitions. He also wrote two pellucid and now classic books that have influenced civilian dressage and equitation since their publication: *Riding and Schooling Horses* (1934) and *Training Hunters Jumpers and Hacks* (1937), the former a guide for developing "a good rider," the latter for educating that rider in "breaking and training the horse." Chamberlin argued in them that "modern warfare has brought a demand for extreme mobility in a cavalry," but that such mobility demands formal dressage, or "proper gymnastic exercises . . . to produce obedience, balance and suppleness," not as an end, but "as a means of making the horse obedient and clever at his normal work." Elaborating that argument, George Morris observes, "Not only was [Chamberlin] a great rider . . . but he was the consummate horseman."[3]

More than a century before Chamberlin, the brilliant cavalryman Friedrich von Bismarck, making a distinction with a difference, had acknowledged the fundamental importance of military horsemanship to tactics, though only as "the foundation of the edifice of tactics; but no more tactics, than a foundation is a building." If horsemanship was the foundation of tactics, the horse was the foundation of horsemanship. Drawing an analogy from artillery, Bismarck identified "horses, horse appointments, and arms [as] the *matériel* of cavalry; and the men . . . the *personnel*," and then went on to observe, "How much more importance the cavalry itself attaches to the *matériel* than to the *personnel*, may be seen from this, that its strength is never reckoned, as in infantry, by the number of men, but by the number of horses."[4] The science of horsemanship trained the horse soldier to exploit fully the horse's fundamental asset, mobility, enabling the military horse to play its defining role. Yet the horse, of course, provided horsemanship its raison d'être. No foot, no horse; no horse, no horsemen, nor horse soldiers, nor mounted warfare. The whole enterprise rests on the foundation of the horse: a perfectly obvious truth, but one worth underscoring.

Viewed from another angle, the horse was the foundation of over four centuries of mounted warfare, but untrained or poorly trained horses ridden by unskilled or marginally skilled riders, it is obvious, would have proven not only dangerous, but also useless, to cavalries. Officers and the horse breakers and riding masters under their command, then, as a matter of practicality and sustainability, not only had to train horses and men, but also had to take men to a level of horsemanship that would allow *them* to train horses.[5] To learn current theories and practices needed to accomplish their mission, and to understand that mission in the context of the evolving history of mounted warfare, cavalry officers depended on authoritative works on military horses and horsemanship and on cavalry arms and organization. If only for that reason, and there are many others, the history of modern mounted warfare in Europe evolved hand in glove with the historiography of mounted warfare, including the primary works considered in the preceding study.

A book about primary works should end with an illustration rather than an exposition, in this case, Jean Renoir's sublime elegy, *La Grande Illusion* (1937), a film about the cavalry of the clouds in the Great War.[6] While *Wings* (1927), *Hell's Angels* (1930), *The Dawn Patrol* (1930, remade 1938), and other examples of the genre derive their intensity from their depiction of aviators *in* combat, *Grand Illusion* (to use the English title) derives its eloquence from its depiction of aviators removed *from* combat—POWs shot down, captured, and imprisoned. An "escape" film, as Renoir later described it, *Grand Illusion* represents a true classic in both its elegant conception and subtle execution. Unlike Hemingway and Sassoon, Renoir (1894–1979) pointedly uses the specific, concrete, and individual to plumb the general, abstract, and universal. *Grand Illusion*, at its highest level, offers a lucid and rigorous poetic essay on affinities between human beings, on the divides and boundaries that destroy or impede those affinities, and on the potential offered by duty, solidarity, and love to affirm them once again, even if such "universals," in the end, may themselves be illusory.

Highly symmetrical in form, *Grand Illusion* opens on two parallel scenes that introduce three main characters: one, a mess for French officers, where Maréchal is assigned to take a staff officer, de Boeldieu, on a surveillance flight; and two, a mess for German officers, where a staff officer, von Rauffenstein, who has shot down Maréchal and de Boeldieu, invites the two pilots to lunch. The body of the film unfolds in two parallel locations: first, a prison camp, where Maréchal, de Boeldieu, and other French POW officers (including Rosenthal, a fourth main character) plan an escape thwarted by their transfer to another POW camp; and second, a fortress-turned-prison, where Maréchal, de Boeldieu, and Rosenthal, each having attempted multiple escapes at other camps, find themselves prisoner, and where von Rauffenstein, since shot down and badly burned and injured, serves as prison commandant. The film reaches its climax, first, with de Boeldieu's sacrificial death at von Rauffenstein's reluctant hand, and second, with Maréchal and Rosenthal's successful escape; and it finds its denouement and end in a final pair of parallel actions: one, after Elsa, a young German war widow and the fifth main character, shelters and hides Maréchal and Rosenthal, she and Maréchal fall in love; and two, Maréchal and Rosenthal, leaving Elsa's home and evading German patrols, reach the Swiss border.

Grand Illusion diagrams and explores the intense and complex relationships among those characters primarily in fluid but juxtaposed pairings— relationships that, on the one hand, are mutable and potentially boundless while, on the other hand, are fixed and constrained by the intersecting boundaries of social class and nationality. In the relationship most pertinent to this study, de Boeldieu and von Rauffenstein, separated by their nationalities, share an assumed mutual respect and understanding based on their social class as aristocrats, military status as career officers, and unwavering commitment to "smartness" and duty. Exploiting an inherent feature of sound film, Renoir makes spoken language the principal metaphor for the multiple boundaries in play. Though de Boeldieu and the working class Maréchal, for example, are native francophones, they speak different "languages."[7] By contrast, de Boeldieu and von Rauffenstein, in addition to their native languages, both speak a fluent and colloquial English that not only marks them as members of the cosmopolitan European ruling class but also allows them to speak privately in the presence of others and familiarly when alone.[8] Language, as much as duty, binds them.

Something else binds de Boeldieu and von Rauffenstein: they are accomplished horsemen. In their first meeting, von Rauffenstein learns that a Comte de Boeldieu, a former acquaintance in Berlin, is de Boeldieu's cousin: "He was a marvelous rider," von Rauffenstein recalls in English. "Yes, in the good old days," de Boeldieu responds, also in English. Much later, in a pivotal tête-à-tête in the fortress, von Rauffenstein, again asking after the Comte, learns that he has lost an arm: "I am so sorry. . . . Such a good horseman!" This, in turn, leads de Boeldieu to recall the mare on whom von Rauffenstein "won the Grand Military at Liverpool in 1909." And this leads to von Rauffenstein's eulogy to an aristocratic social and military order: "I do not know who is going to win this war, but I know one thing: the end of it, whatever it may be, will be the end of the Rauffensteins and the Boeldieus." Demurring on the "pity" of that, de Boeldieu foreshadows his subsequent self-sacrifice, one that will force von Rauffenstein, with tragic inevitability, to shoot him, but also will enable Maréchal and Rosenthal to escape. Dying, true to "smartness" and with a light hand, de Boeldieu soothes von Rauffenstein's piercing remorse: "I would have done the same thing. French or German . . . duty is duty."[9]

Renoir condenses *Grand Illusion* into that statement and its unspoken equivalents in the escape and final two scenes: for Maréchal and Rosenthal, solidarity is solidarity, and, for Maréchal and Elsa, love is love. Renoir builds a compelling humanist vision, in a word, on duty, solidarity, and love. When Maréchal says early in the film, however, that the war soon will be over, he is told, "You're under an illusion"; and when he wearily says, in the film's closing shots, "We've got to finish this bloody war. . . . Let's hope it's the last," it is Rosenthal, in Renoir's grim irony, who replies, "That's all an illusion!"[10] De Boeldieu dies in the war; von Rauffenstein, "not finished dragging out a useless existence . . . has missed" escape through death; Maréchal and Rosenthal cross the "man-made frontier" invisible beneath the snow, but only in order "to fight again." Though humanist and poetic, Renoir's vision is not ethereal. The film dramatizes some escapes to suggest that others are possible, just as it dramatizes individual human connections to suggest that collective human connections are possible. It also foresees, however, that such optimism, in the end, may be the grandest illusion.

In 1958, Renoir introduced a restoration of *La Grande Illusion* with a filmed statement that looks back not just to his film set in the Great War and released in 1937, but also to the rise of Nazism between 1918 and 1937—an epitome of barbarism whose eventual fall in 1945 had exacted the *next great war* and the millions of lives lost in it. "In 1914," Renoir says, "Hitler had not yet appeared. Nor had the Nazis, who almost succeeded in making people forget that the Germans are also human beings. In 1914, men's spirits had not yet been warped by totalitarian religions and racism. In certain ways, that world war was still a war of formal people, of educated people—I would almost dare say, a gentlemen's war. That does not excuse it. Politeness, even chivalry, does not excuse massacre." A year before *La Grande Illusion* first appeared, Siegfried Sassoon had grieved bitterly for the Great War victims, especially the shell-shocked, "who in the name of righteousness had been sent out to maim and slaughter their fellow-men. In the name of civilization these soldiers had been martyred, and it remained for civilization to prove that their martyrdom wasn't a dirty swindle."[11] Sent out by Renoir's gentlemen? or by Sassoon's swindlers? Only the dead can tell.

Notes

Preface

1. Bismarck. *Lectures on the Tactics of Cavalry,* 278; Tomassini, *The Italian Tradition of Equestrian Art,* 186–87.

2. For definitions of heavy cavalry, light cavalry, and dragoons, see "Cavalry Terms" in chapter 3.

3. Littauer, *The Development of Modern Riding,* 127.

4. Rodzianko, *Modern Horsemanship,* 154–55.

5. Kenyon, *Horsemen in No Man's Land,* 242.

6. The "long nineteenth century" generally refers to the period framed by the Napoleonic Wars of 1796–1815 and the Great War of 1914–1918 (see, for example, Coulston, introduction to Nolan, *CHT,* xiv).

7. DiMarco, *War Horse,* 350.

8. Podeschi, *Books on the Horse and Horsemanship,* ix; Van der Horst, *Great Books on Horsemanship,* 18.

9. Nomenclature was fluid in horsemanship as in any intellectual discipline over this period. The noun *horsemanship* first entered English in Thomas Blundeville's *The fower chiefyst offices belongyng to Horsemanshippe* (1565–1566), as a general term pertaining to use and care of horses. (Editions of Blundeville's work following the second edition of 1570 appeared as *The Foure Chiefest Offices belonging to Horsemanship.*) The term horsemanship took on its more specific application to dressage and equitation only much later, when horsemanship as skill in training and riding was joined by the term *horsemastership* to mean skill in the management and care of horses. (The latter term dates to the nineteenth century, as indicated by the *Oxford English Dictionary,* when it also entered common military usage.) Concurrently, over this period, the term *farriery,* encompassing horse care and particularly the treatment and cure of horse ailments and injuries, evolved into its more specific application to shoeing and foot care in the eighteenth century, when *veterinary medicine,* as a discipline and a term, was formalized and institutionalized.

10. The one exception in early texts is the substitution of <s> for the medial or long <s> that resembles an <f> to modern eyes.

1. Ryding and Breakinge

1. Tomassini, *The Italian Tradition of Equestrian Art*, 215; Littauer, *The Development of Modern Riding*, 77.

2. Tomassini, *The Italian Tradition of Equestrian Art*, 217.

3. Morgan, preface to Xenophon, *The Art of Horsemanship*, 37. Not Xenophon's first English translator, Morgan found earlier versions, including Richard Berenger's translation in *The History and Art of Horsemanship*, (1771), "unsatisfactory" and not of "much assistance" (Morgan, preface to Xenophon, *The Art of Horsemanship*, i).

4. Littauer, *The Development of Modern Riding*, 15; Felton, *Masters of Equitation*, 17; Morgan, preface to *The Art of Horsemanship*, 51.

5. Morgan, preface to *The Art of Horsemanship*, 40; Littauer, *The Development of Modern Riding*, 18; Morgan, preface to *The Art of Horsemanship*, 51.

6. Xenophon, *The Art of Horsemanship*, 8; 25. With regard to obedience, Xenophon cautioned, "A disobedient servant is of course a useless thing, and so is a disobedient army; a disobedient horse is not only useless, but he often plays the part of a very traitor" (*The Art of Horsemanship*, 7).

7. Xenophon, *The Art of Horsemanship*, 11; 21; 17–19; 14. "Those who are enlisted in the cavalry in our states are persons of very considerable means, and take no small part in the government," Xenophon writes, so they cannot be expected to break colts (*The Art of Horsemanship*, 5).

8. Xenophon, *The Art of Horsemanship*, 34.

9. Xenophon, *The Cavalry General*, unpaginated. Xenophon also anticipates *The Art of Horsemanship* when he emphasizes the need for horses with good feet who "will stand being ridden over rough ground," and for troopers who can "ride with freedom" and "keep a firm seat" over "every sort of ground."

10. Reflecting the nexus of military leadership and artistic creativity, the same word—*auctor*—applied to both generals and authors. The experimental American writer, Gertrude Stein, would make much of that double meaning in *Four in America* (1947), her nearly opaque biographical "studies" in militarism and creativity.

11. Felton, *Masters of Equitation*, 19; Chenevix-Trench, *A History of Horsemanship*, 115. Littauer opines that "educated riding—in the strict sense of the term—did not exist before the 16th century or, if it did, we have no knowledge of it" (Littauer, *The Development of Modern Riding*, 50); but Tomassini writes that "even if there weren't printed works on equitation before the publication of Grisone's treatise in 1550, we know that horsemen often passed on their 'secrets' to the narrow circle of their disciples in manuscript form" (Tomassini, *The Italian Tradition of Equestrian Art*, 55).

12. On armor, see Pyhrr et al., *The Armored Horse in Europe 1480–1620* (2005). On equines, see Sir Walter Gilbey, *The Great Horse, or The War Horse: from the time of the Roman Invasion till its development into the Shire Horse* (1889). As

Felton has observed: "But all of this armor did not lead to high standards of equitation. Agility could not be expected and the armor itself must have made impossible even a crude use of the legs. About the only test of horsemanship was the
rider's ability to stick on" (Felton, *Masters of Equitation,* 21).

13. Dom Duarte's treatise, as Tomassini recounts, "has come to us through a
manuscript, written around 1434, acquired by the Royal Library of Paris, under
Colbert, subsequently copied in 1830 and first published in Paris in 1842"
(Tomassini, *The Italian Tradition of Equestrian Art,* 58). The only currently available English translation of Dom Duarte has so many errors in syntax, grammar,
and diction that I correct them silently when quoting the text.

14. Dom Duarte, *The Art of Riding,* 41 and passim.

15. Tomassini, *The Italian Tradition of Equestrian Art,* 40; Dom Duarte, *The
Art of Riding,* 44, 185, 134. Dom Duarte regards reading about riding unnecessary, but still "beneficial" (Dom Duarte, 43). He also anticipates many future
authorial claims and disclaimers when he adds that he bases his conclusions on
"my experiences—more than from reason" (120); strives to achieve clarity rather
than elegance in writing (130); and has "been practicing and learning this science
well—namely, the art of being a good horseman; and this is the reason why I have
decided to write about it" (213). He also echoes Xenophon and anticipates later
writers when he advocates for hunting as ideal training for cavalry, though, unlike
most later writers, he also reverses the hunting equation, arguing that training for
war also makes one a better hunter in peacetime.

16. Dom Duarte, *The Art of Riding,* 78, and see 70, 124, 164.

17. Dom Duarte, *The Art of Riding,* 114–15, 117, 158.

18. Dom Duarte, *The Art of Riding,* 95, 81–82.

19. Dom Duarte, *The Art of Riding,* 109. The translator's gloss on Dom Duarte's point about quietness and tranquility—"we can consider the quietness of the
body as a physical consequence of the tranquility the horseman feels" (109n)—is
accurate but inadequate, since Dom Duarte also is gesturing toward the converse
idea that mental tranquility is a consequence of physical quietness.

20. Dom Duarte, *The Art of Riding,* 59, 102, 67–68.

21. Biographical details on Grisone are scant. He was born in the late fif-
teenth century and died in the late sixteenth century, but the exact dates of his
birth and death are unknown (see Tobey, introduction to Grisone, *The Rules of
Riding,* 34–40; and Tomassini, *The Italian Tradition of Equestrian Art,* 79–84).
Tomassini and Tobey both imply that Grisone most likely did *not* know Dom
Duarte's manuscript (Tomassini, 58–60; Tobey, 27–28) but *might* have known
Xenophon's printed treatise. Tomassini cites the publication of Xenophon in
Greek in Florence in 1516 and in Italian translation in Venice in 1580 (Tomassini,
49–50). Noting, in addition, a Latin translation of Xenophon in 1539 and a
Spanish translation in 1552, Tobey allows that Grisone "may have known about
Xenophon's treatise and might even have read it," but believes that, even if so, "it
had only a marginal influence on the *Ordini* (Tobey, 28–29). For bibliographical

details on the 1580 Italian translation of Xenophon's *On Horsemanship,* see van der Horst, *Great Books on Horsemanship,* 200–201.

22. Tomassini, *The Italian Tradition of Equestrian Art,* 79. Tomassini reports that "between 1550 and 1623, there were twenty printed Italian editions, fifteen French translations, six English, seven German and one in Spanish" (Tomassini, 84). Citing slightly different figures, Tobey writes that "twenty-three separate printed editions of the *Ordini* appeared in print in the original Italian between 1550 and 1620," that English, French, German, and Spanish translations appeared within the first two decades of initial publication, and that, "between 1559 and 1623, twenty-eight separate editions of translations . . . were printed" (Tobey, introduction to Grisone, *The Rules of Riding,* 5, 13).

23. Chenevix-Trench, *A History of Horsemanship,* 101; Tobey, introduction to Grisone, *The Rules of Riding,* 1, 40. The few *printed* works on horses that preceded the *Ordini* primarily "concern veterinary medicine and animal husbandry" (Tobey, 25) and "mouthpieces" (Tomassini, *The Italian Tradition of Equestrian Art,* 57), not dressage or equitation. Tomassini offers two arguments mitigating claims for Grisone's originality. First, the shift in the sixteenth century from manuscript to print culture benefited Grisone: the *Ordini* in print circulated on a different scale than did prior texts in manuscript, so, whether original or not, it "took on the value of founding a new tradition of works dedicated to the horse and its riding." Second, claims for Grisone's originality serve to "minimize the brutality of Grisone's methods [by] emphasizing his role as beginner of the new literary genre of treatises on horsemanship" (Tomassini, 17, 99).

24. Grisone, *The Rules of Riding,* 67; Guy, preface to *The Rules of Riding,* xxi. Guy notes that Grisone discussed, among other matters, rhythm, relaxation, contact, impulsion, straightness, and collection. See also Tobey's discussion of the concept of "contact" in the *Ordini* (Tobey, introduction to Grisone, *The Rules of Riding,* 52–57).

25. Grisone, *The Rules of Riding,* 393.

26. Grisone, *The Rules of Riding,* 63, 65, 175. As Tobey points out, Grisone was "the first author to use the term *aere* or *aria,* 'air'—a term borrowed from Renaissance dance—in the context of horsemanship to describe the graceful vertical movement of the horse's body" (Tobey, introduction to Grisone, *The Rules of Riding,* 48). As Tomassini notes, however, it was Cesare Fiaschi who brought "to the extreme that 'musical' conception of horseback riding that puts the notions of time and measure as the foundation of the equestrian art." As a result, "in Fiaschi's view of horsemanship, there appears a more marked distinction [between training for war and training as art]" (Tomassini, *The Italian Tradition of Equestrian Art,* 109, 124).

27. Chenevix-Trench, *A History of Horsemanship,* 103; Tobey, introduction to Grisone, *The Rules of Riding,* 30, 42; Tomassini, *The Italian Tradition of Equestrian Art,* 43, 47. With respect to martial application, Littauer observes that when high school movements were introduced in the sixteenth century, "some of their

sponsors liked to claim that these were practical for war or tournaments. I suspect that they made these claims simply to sell their methods to the hard-headed practical [noble]man." Shifting from riding masters to their patrons, he similarly observes, "What High School did was to take the prancings and rearings and cavortings, to label and regularize and control them, and put them into a system to serve the purpose of the age—the glorification of the great" (Littauer, *The Development of Modern Riding,* 54, 63).

28. On those points, Tobey notes that Grisone's emphasis on measure and time, representing Renaissance values of proportion and moderation, suggests the degree to which he "reflects some of the central philosophical ideas of his era" (Tobey, introduction to Grisone, *The Rules of Riding,* 33–34); and Tomassini writes: "Fully consistent with classical aesthetics that informed the Renaissance, riding is sublimated in the supreme paradox: using the maximum artifice to attain [or simulate] maximum naturalness" (Tomassini, *The Italian Tradition of Equestrian Art,* 42).

29. Since an educated horse, of his own will, "becomes one with the will of the rider who sits upon him," the rider "must ride and sit upon the horse, not only with great heart without fearing him, but also envisioning that you and he are one as the same body, feeling, and will." To the spectator, it follows, "it will seem that he and you are of one body, of one sentiment, and of one will" (Grisone, *The Rules of Riding,* 99–101, 107, 399). With respect to the horse's nobility, Grisone, virtually echoing Job 39:19–39:25, asks: "Now, who could ever fully tell you the great praise and great virtues of the Horse? Who does not recognize the horse as king of the animals . . . and the most faithful companion of kings?" (Grisone, 67). In eliciting the innate nobility of the horse, of course, the rider reveals his own innate nobility.

30. Brown, introduction to *Virtue and Beauty,* 13; Grisone, *The Rules of Riding,* 97–99; Tobey, introduction to Grisone, *The Rules of Riding,* 34.

31. Felton, for example, writes that the method of Grisone and his peers, "based on complete subjugation of the horse," was designed to produce horses "so completely under control as to be the creatures of the rider's will" (Felton, *Masters of Equitation,* 26–27). Littauer observes that the "cruelty" often ascribed to Grisone and his peers "stemmed not only from the general cruelty of the times but also, ironically, from an attitude towards the horse that credited him with a more human type of mentality than we ascribe to him today" (Littauer, *The Development of Modern Riding,* 71). The general anthropomorphosis of the horse, in short, resulted in attributing to the equine not only human virtues, such as kindness, courage, and loyalty, but also human vices, particularly falseness and maliciousness.

32. Grisone, *The Rules of Riding,* 349. Grisone recommends dealing with laziness, even more egregiously, not only by jabbing the horse "continually with the spurs . . . and so frequently that it draws blood near the girth," but also by repeating the action the following morning, when "he will feel the jabs with the spurs more acutely" because the still fresh wounds will be "painful and cold" (Grisone, 239). He also makes clear, though, "that a rider of good discipline will not ever avail himself of these [or similar] methods, since he can have the same results

without these methods, using his own virtue instead in different ways" (Grisone, 347–49).

33. Grisone contends that when a horse balks, "most of the times this is the fault of the rider" (Grisone, *The Rules of Riding*, 335), and "when the horse pulls on the bit and runs away, [this opposite vice] arises not only from a bad temperament but from the training that the rider has given him, which was without reason and order" (Grisone, 351). He also advises, "Each time that the horse is ridden, leave him in good spirit" (177). Indeed, the rider should never "weaken [the horse's] spirit, because in the end, with mindfulness, you will find that he will very easily do whatever small thing you ask of him" (393–95). Grisone's entire system, like most systems of dressage, is based on aid *and* correction, reward *and* punishment. As often as Grisone recommends hitting a horse with a crop or stick (see 105, 347, or 353), he admonishes stopping as soon as the horse responds (see 135 or 357), and he encourages reassuring the horse with a pat of the hand (see 101, 105, 135, or 363). As often as he recommends using a harsh and loud voice to threaten a horse (see 105 or 343), he encourages "reassuring him with a deep and pleasant voice" (217; also 363, 397).

34. Grisone, *The Rules of Riding*, 357, 397.

35. Tobey, introduction to Grisone, *The Rules of Riding*, 43; Felton, *Masters of Equitation*, 43; Tobey, 18; van der Horst, *Great Books on Horsemanship*, 128; Guy, preface to Grisone, *The Rules of Riding*, xv.

36. Guy, preface to Grisone, *The Rules of Riding*, xv; Tobey, introduction to Grisone, *The Rules of Riding*, 19; Blundeville, *The Arte of Ryding and Breakinge Greate Horses*, numbered but unpaginated plates. The full title of Blundeville's work, though rarely used, is *A Newe Booke containing the Arte of Ryding and Breakinge Great Horses* (etc).

37. Tobey, introduction to Grisone, *The Rules of Riding*, 18–22; Blundeville, *The Arte of Ryding* (1597), 35–36. This advice, also quoted by Tobey (Tobey, 22) and Felton (*Masters of Equitation*, 47), does not appear in the 1560 edition of *The Arte of Ryding*, though I have seen it in subsequent early editions.

38. Blundeville, epistle to *The Foure Chiefest Offices belonging to Horsemanship*, n.p. Blundeville had dedicated *The Arte of Ryding* to "L. Roberte Dudley, Knight of the honorable order of the garter, & master of the Queens highnes horses." Dudley subsequently was created Earle of Leicester and Baron of Denbigh, and is identified as such in the dedicatory epistle to *The Foure Chiefest Offices*. All citations to *The Foure Chiefest Offices* are from the edition of 1597.

39. Blundeville, epistle to *The Foure Chiefest Offices belonging to Horsemanship*, n.p.

40. Blundeville, preface to *The Arte of Ryding*, n.p. Astley returned the compliment in the dedication to his *Art of Riding*, citing Blundeville's *Arte of Ryding* as such a skillful translation and adaptation of Grisone's work that "as if men take good heed, & wilbe diligent, they cannot but greatlie profit thereby, to the great benefit of themselues, and the seruice of their countrie" (Astley, dedication to *The Art of Riding*, n.p.).

41. Van der Horst, *Great Books on Horsemanship,* 132. Astley dedicated *The Art of Riding,* "To the Right worshipfull Gentlemen Pensioners, M. Henrie Mackwilliam, and M. William Fitzwilliams" (n.p.) who, in turn, added their own dedicatory note, "To our verie louing Companions, and fellowes in Armes, her Maiesties Gentlemen Pensioners." In it, they characterize Astley as "a man . . . knowne to be of singular skill in the Art of Riding" and "the onelie man, that persuaded Maister Blundeuill to take first in hend his worke of Frederike Gryson" (Mackwilliam and Fitzwilliams, n.p.).

42. Biographical material adapted from van der Horst, *Great Books on Horsemanship,* 128, 132.

43. Astley, "letter missiue," to *The Art of Riding,* n.p.; Astley, *The Art of Riding,* 1, 2, 6, 16, 22, 39, 42, 52; Astley, *The Art of Riding,* 11–12. Since Xenophon and Grisone have "litle or nothing" to say about proper use of the "Cauezzan" (cavesson) as Astley notes (20), he appends to his text a "discourse . . . of the *Chaine* or *Cauezzan,* and likewise of the Trench & Martingale . . . not the Authors worke, but the experience of another Gentleman" (Astley, 69; text of "discourse," 69–79).

44. Astley, dedication to *The Art of Riding,* n.p., and see 2. Given Astley's emphasis on horsemanship for military service, he chooses an apt analogy: "For how shall one make another vnderstand, to what purpose the pomell of a sword serueth, if he shew him not first what a sword it selfe is?" (Astley, dedication to *The Art of Riding,* n.p.).

45. Tobey writes that Astley was "perhaps the master who best understood Grisone's theories of contact," called *appoggio* by Grisone and translated as *appuy* by Blundeville. Tobey also offers convincing evidence that Astley read Grisone in Italian as well as Blundeville in English (Tobey, introduction to Grisone, *The Rules of Riding,* 55).

46. Astley, *The Art of Riding,* 16, repeated almost verbatim 42; Astley, 39.

47. Astley, *The Art of Riding,* see 5, 3, 44, 3, 8. Astley follows Grisone (and Blundeville) on most other critical points, including, for example, the need for the rider to "accompanie [his horse] in time and measure, so as to the beholders it shall appeare, that [horse and rider] be one bodie, of one mind, and of one will" (Astley, 57).

48. Astley, *The Art of Riding,* 1, 4. When Astley finally does speak of military equitation, he turns the question of "the right use of the hand [for] the teaching and making of a horsse" to that of "the vse of the hand vpon a horsse alreadie taught, and fit for the seruice, wherein we haue but the vse of the left hand onelie: for the other must serue vs for our weapon whatsoeuer it be" (Astley, 52). He closes, in short, on a point that would echo through the next three centuries of writing on military equitation, namely, that riders engaged in combat had to use only their left (bridle) hand to control the horse, because they used their right (sword) hand to fight their adversaries.

49. Van der Horst reports, "Together with Grisone, Fiaschi, and Pignatelli, Corte belongs to the most important Italian authors on horsemanship and horse

training, all of whom were invited to enter into the service of various European courts." Corte, to cite the case in point, "had spent a year in England [in 1565] at the invitation of Robert Dudley," and returned to England "in 1573 to become the expert counsellor on the horses and horsemanship at the England court" (van der Horst, *Great Books on Horsemanship*, 132, 190). Tomassini attributes "a considerable homogeneity of the equestrian practice of the time" to the fact that, "moving from court to court, the best horsemen spread the equestrian doctrine in Italy and in Europe" (Tomassini, *The Italian Tradition of Equestrian Art*, 161).

50. Astley, *The Art of Riding*, 58. In the dedicatory epistle to Lord Robert Dudley in *The Foure Chiefest Offices belonging to Horsemanship*, Blundeville speaks of his plan to write "another little Book of Additions . . . to ioine to this volume, briefly comprehending all the precepts of a later writer, now being your Honours most excellent Rider, called master *Claudio Corte*," and including anything that Grisone or Blundeville, in the *Ordini* or its translation, "perhaps haue negligentlie omitted" (Blundeville, epistle to *The Foure Chiefest Offices belonging to Horsemanship*, n.p.).

51. Mackwilliam, "To the right worshipfull, my verie louing Companions, and fellowes in Armes, hir Maicsties Gentlemen Pensioners," *The Art of Riding*, n.p. See also Bedingfield's letter, "To the Reader," *The Art of Riding*, n.p. Bedingfield had established his bona fides with a translation, in 1573, of "a philosophical work by Girolamo Cardano" (van der Horst, *Great Books on Horsemanship*, 194).

52. Tomassini, *The Italian Tradition of Equestrian Art*, 142. Written as a dialogue in imitation of Castiglione's *Book of the Courtier*, book 3, the most consequential in Tomassini's view, intends "to establish the figure of the horseman as a precise social role"—one parallel and equivalent to that of the courtier (Tomassini, 155).

53. Citations are from Bedingfield's letter, "To the Reader," except for Bedingfield's reference to Astley (Bedingfield, 66). For Bedingfield's references to Corte in the third person, see, for example, 17, 50, 90. In "To the Reader," Bedingfield proposes that the art of horsemanship "hath neuer beene (I meane within this realme) of that perfection it now is. . . . For, before M. Blundeuile, I find not anie that haue written in our toong: neither were the teachers of that time of much knowledge" (Bedingfield, n.p.).

54. On affectation as opposed to comeliness in "seat," see Bedingfield, *The Art of Riding*, 33, 35, 36, 50, 73; on gentleness as opposed to violence, see 97, 105.

55. Bedingfield, *The Art of Riding*, n.p.; n.p.; 45; chap. 28, 93–95; 94. "A horse for the warre," Bedingfield concludes, "ought to be a swift and sure runner, a good eater, light vpon the hand, strong, nimble, and valiant, without fault or imperfection" (Bedingfield, 95).

56. Tomassini, *The Italian Tradition of Equestrian Art*, 137. Astley's treatise appeared earlier than Bedingfield's translation, and the two texts appear to have been bound together in that order. I base the generalization on van der Horst's description of the copy in the library of Johan Dejager (van der Horst, *Great Books on Horsemanship*, 190), and on my examination of the copy held in the National Sporting Library & Museum.

57. Tomassini, *The Italian Tradition of Equestrian Art,* 38.

58. I can find no biographical information on Morgan beyond the attribution on the book's title page: "Nicholas Morgan of Crolane, in the Countye of Kent, Gent." The catalogue entry for a copy of *The Perfection of Horse-manship* auctioned at Christie's in 2006 indicates, "The sheets of this work . . . were subsequently reissued in 1620 . . . under the title *The Horse-mans Honour, or, The Beautie of Horsemanship*" (Christie's, *Sale 7300*). In *Books on the Horse and Horsemanship,* Podeschi writes, "After page 16, [*The Horse-mans Honour*] appears to be printed from the same typesetting as [*The Perfection of Horse-manship*] (Podeschi, 26). My examination of the 1609 and 1620 editions of the two works in the holdings of the National Sporting Library & Museum confirms Podeschi's analysis. The preliminaries and pages 1–16 of NSLM's copies differ in contents, chapter headings, running heads, decorated initial capitals, and fonts. Beginning with page 17, the two volumes obviously were printed from the same plates (and may indeed share the same sheets). All aspects are identical, including the several instances of mispagination.

59. In the first, "The Epistle to the Kings Maiestie," Morgan argues that Creation endowed the horse—who is to other beasts what the king is to men—with virtues, "So that the glory of Princes can be by none more highly aduaunced, their Armyes more inuincibly fenced, or their Enimies more speedily subuerted." For this reason, Morgan "haue drawen from the springs of Nature, Arte and Practise" in order to revivify the "now wthiered & dead Art of Horsemanship." The "Skill of Horsemanship . . . newly reuiued," however, as he writes in the second epistle, "To Prince Henry," requires comparable skill in breeding so "that your men shall not complain for want of excellent Horses, nor your Horses groane for want of worthy Riders." This is why, he writes in a third epistle, "To the Earle of Worcester," that he has labored to discover and share not only "the hidden Secretes of Horsemanship, but also the manifest Errors of the Arte and Practises" in current use. The epistles are unpaginated.

60. Morgan's argument unfolds roughly as follows. Creation had endowed humanity with supremacy over the animal kingdom, but the Fall had forfeited that supremacy. After the Fall, though, man's original virtues, though obscured by his "transgression," had remained "planted in his originall Nature, so as he shall euermore desire the true knowledge and practice thereof, because nature still desireth restitution to his prymary perfection" (Morgan, *The Perfection of Horsemanship,* 1–3). Since nature drives humanity to regain its prelapsarian state, and since humanity must use art—"but a hand maid to nature" (Morgan, 345)—as the means to that end, no art can oppose nature and all art must follow nature. Hence, as Morgan explained in "Admonitions to the Reader," "this whole worke taketh his grounds from nature" and "the true secrets of Nature" (Morgan, n.p.).

61. Morgan, *The Perfection of Horsemanship,* 2, 6, 9.

62. Morgan, *The Perfection of Horsemanship,* 89–93, 94; 215–16.

63. In brief summary, Morgan begins with character: "A good Horseman and perfet Rider, must not onely haue naturall gifts of true valoure, wisedome and temperance, but also true knowledge and practise to attain perfection . . . whereof the ignorant & pretended Rider proceedeth to violence, which nature abhorreth, as Arte doth error and reason vnruly passion" (Morgan, *The Perfection of Horsemanship*, 166–68). He argues that knowing how and *when* to aid, correct, or praise are "the onelye and principall thinges required in a perfect Ryder" (Morgan, 167), and that the primary tools for them are "the voyce, the hand, and the legge"—the hand being "the instrument of instruments in the true vse and gouernment whereof is the ground of the whole Art" (170, 172). Finally, "In all his dooinges, from the beginning to the end, [the rider must] keepe his reine true, and his [horse's] head steady, for it [contact] is the foundation of all" (213).

64. Morgan, *The Perfection of Horsemanship*, 42, 346.

65. Markham's first book on horsemanship, *A Discourse on Horsemanshippe* (1593), reedited and retitled in 1595, had four additional editions prior to 1606 and led to the publication of *Cavelarice* in 1607. *Markham's Maister-peece*, first published in 1610, "was re-edited at least 20 times between 1615 and 1734" and published in French translation in 1666. A corrected and enlarged twelfth edition in 1681 was the first to carry an appended work, *The Complete Jockey; Containing Methods for the Training of Horses up for Racing*. An abridged edition for farriers, *Markham's Faithful Farrier* (1630), enjoyed new editions through 1883 (see van der Horst, *Great Books on Horsemanship*, 294–95, 299.)

66. Steggle, "Markham, Gervase," *Oxford Dictionary of National Biography*. Though "one of the most prolific and popular English writers of his day," Markham suffered both contemporary and later detractors (van der Horst, *Great Books on Horsemanship*, 294). John Lawrence devoted eight pages of his *Philosophical and Practical Treatise on Horses* (1802) to denouncing "the redoubtable Gervase Markham, for more than a century, the oracle of sapient grooms, the fiddle of old wives, and the glory of booksellers. . . . He was, in my opinion, nothing better than a mere vulgar and illiterate compiler; and his works . . . are stuffed with all the execrable trash that had ever been invented by any writer . . . on the subject of horses" (Lawrence, 9–10).

67. *Cheape and Good Husbandry* comprises two books: the first, *Of Beasts*, includes a long section on the horse, followed by five shorter sections on other farm animals; the second, *Of Poultry*, includes four sections on farm and wild poultry, plus sections on bees and on fishing. The section on the horse occupies nearly half the volume. Preliminaries to *Cheape and Good Husbandry* are unpaginated.

68. Markham, *Cheape and Good Husbandry*, 2, 3–5, 5–9, 10–41.

69. Markham's points include these: the horseman not only should ride horses perfected by others but also should train young horses himself, thus benefiting "his minde" and "his body" (Markham, *Cheape and Good Husbandry*, 11); "the three maine points of a Horsemans knowledge . . . are helps, corrections, and cherishings" (15); "every Horse naturally desireth neither offence, nor to offend;

but the rash discretion of ignorant Horse-men . . . is the begetting of those evils which are hardly or ever reclaimed" (30).

70. Markham, *Cheape and Good Husbandry,* 31, 34, 38, 35, n.p.

71. A second edition of Hope's full translation appeared in 1717. Van der Horst notes, "An abridged edition was also published in 1696 with the title, *The compleat horseman, or perfect farrier;* with re-editions in 1702, 1717 and 1729" (van der Horst, *Great Books on Horsemanship,* 402). Hope had undertaken the abridgment, he writes in the preface to the edition of 1702, so that "Those who grudge the Price of the Original, or are scar'd by its length, may here gratifie their Curiosity, without any considerable loss either of Money or Time" (Hope, preface to Sollysell, *The Compleat Horseman,* n.p.).

72. Van der Horst, *Great Books on Horsemanship,* 394.

73. A fourth point concerns the story of the book's organization. In "The Authors Epistle To the Reader" of the eighth French edition (translated and included by Hope), Solleysell decries the many counterfeit (that is, "pirated") editions of his work. So future readers can identify them, he now has changed the order of his prior authorized editions, reversing parts 1 and 2. Noting the change and its reason, Hope reverses the two parts back to their original order. Were this not sufficiently baroque, "An Advertisement by the Publisher," prefacing what was now part 2, explains that Hope and a Scots writer, unbeknownst to one another, were translating Solleysel concurrently, but in a different order, creating a confusion of section titles and numbers (both are labeled part 1) that takes the publisher two full pages to sort out. Preliminaries to *The Compleat Horseman* are unpaginated, as is "An Advertisement by the Publisher."

74. Following an unacceptable French translation of Newcastle's *A New Method, and Extraordinary Invention, to Dress Horses* (1667) that had appeared in 1671, Solleysel undertook a second translation, with Newcastle's approval, that appeared in 1677.

75. The publisher's advertisement boasts that the addition of Hope's *Supplement* means that "any Gentleman may be made, by the serious Perusal of [part 1 of] this Book, not only a *Skilful* and *Compleat Farrier,* but also an *Understanding* and *Compleat Horseman*" (Solleysel, *The Compleat Horseman,* n.p.).

76. Solleysel, *The Compleat Horseman,* 1696, 1; and *The Compleat Horseman,* 1702, 145.

77. "But in the science or art of equitation," Chenevix-Trench writes, "one can surely see a clear distinction between the sixteenth and seventeenth centuries' views. [Horsemen] were, in short, more scientific, and more humane, than Grisone, Pignatelli, and their sixteenth-century English plagiarists" (Chenevix-Trench, *A History of Horsemanship,* 122).

78. Books by early masters differ not only in views on horsemanship from their time to ours, but also in fashions of publishing. Early books, whatever the subject, often had colorful, and sometimes louche, histories peopled by authors, patrons and executors, editors and translators, engravers and printers, booksellers

and book pirates. Texts could vary considerably from edition to edition, even from copy to copy, as could illustrations. The works of de Pluvinel, Cavendish, and de la Guérinière are no exception: each work has a complex publishing and textual history discussed in detail by van der Horst in *Horseman as Bookman*.

79. De Pluvinel, *Le Maneige Royal*, 40. In chapter 1 of part 2 of *École de Cavalerie*, de la Guérinière writes, "Among many authors, we have, according to the unanimous sentiment of all connoisseurs, only two whose works are esteemed, these being M. de la Broue and the Duke of Newcastle" (de la Guérinière, *École de Cavalerie*, 76). In addition to invoking the two masters individually on numerous occasions, de la Guérinière also invokes them as a pair at least two more times (87, 136).

80. Littauer, *The Development of Modern Riding*, 38; and see Felton, *Masters of Equitation*, 25.

81. The edition of 1625 now generally goes by the name of that of 1623. Both van der Horst and Nelson have addressed the complicated publishing and textual history of *Le Maneige Royal* and *L'Instruction du Roy*, including the role of de Pluvinel's "disciple and literary executor," René de Menou de Charnizay, in "additions, which are mostly social commentaries," and the roles of the publishers, Crispen de Pas, the elder, and J. D. Peyrol, and the engraver, Crispin de Pas, the younger, in the textual variants in the early editions (van der Horst, *Great Books on Horsemanship*, 354–57; Nelson, preface to *Le Maneige Royal*, v–vi).

82. De Pluvinel, *Le Maneige Royal*, 15. Cavendish puts a keener edge on the same idea: "If princes were as industrious to know the capacities of men for the different trusts they put in them, as good horsemen are to employ each horse in that which nature design'd him for, kings would be better serv'd than they are; and we should not see such confusion, as surpasses that of BABEL, happen in states through the incapacity of persons entrusted" (Cavendish, *A General System of Horsemanship*, 18).

83. Nelson, introduction to *Le Maneige Royal*, 10; de Pluvinel, *Le Maneige Royal*, 16, 39.

84. De Pluvinel, *Le Maneige Royal*, 24, 21, 31, 39, 43, 54, 93, 94, 102.

85. Cavendish was to the manner born: his uncle, William Cavendish, 1st Earl of Devonshire (1551–1626), the subject of Edwards, *Horses and the Aristocratic Lifestyle in Early Modern England* (2018), was a noted horseman and holder of vast estates who maintained a stud that bred high quality saddle horses for a market of his titled peers.

86. Cavendish, cited in van der Horst, *Great Books on Horsemanship*, 312; de la Guérinière, *École de Cavalerie*, 78. Cavendish's influence extended well beyond the eighteenth century, as E. Schmit-Jensen notes in his technical commentary to *A General System of Horsemanship*: "A century later the great French masters d'Aure and Baucher were continuing to quote from Newcastle in their works, and the German equestrian genius Gustav Steinbrecht regarded Newcastle's book as

the most important in the literature" (Schmit-Jensen, technical commentary to *A General System of Horsemanship*, n.p.).

87. Cavendish, *A General System of Horsemanship*, see 99–101; 105; 105, and see 13; 101; 27, repeated on 131; 142. Cavendish speaks often of spurs and bits. Regarding spurs, he writes, "If the rider begins again to beat and spur, the horse will resist again: it is not the beast then that is vanquished, but the man, who is the greater brute of the two: the whip and the spur serve only to continue the quarrel even to death, as in a duel" (Cavendish, 105). Regarding bits, he advises, "But above all, this rule is chiefly to be observed, to put as little iron in your horse's mouth as possibly you can" (130). Cavendish invokes Pignatelli and de Pluvinel, who broke decisively with earlier masters on bitting, as his authorities on this matter (131). See de Pluvinel, *Le Maneige Royal*, 150; and de la Guérinière, *École de Cavalerie*, 51.

88. Cavendish, *A General System of Horsemanship*, see 17–23; 33 and 63; 68. Cavendish explains, "Thus you see that leaping-horses are disposed by nature, and not art, being full of spirit, and light; so that a horseman hath nothing to do in making leaping-horses, but only to give them the time, which is all the art ought to be used to a leaping-horse; and he that thinks to shew more art in a leaping-horse, will but shew his ignorance and folly" (Cavendish, 70).

89. Cavendish, *A General System of Horsemanship*, 11, and see 58, 122; 33–34, and see 95–97.

90. Tomassini, *The Italian Tradition of Equestrian Art*, 229; de la Guérinière, *École de Cavalerie*, 1.

91. De la Guérinière, *École de Cavalerie*, 1–2; Boucher, preface to de la Guérinière, *École de Cavalerie*, xvi.

92. De la Guérinière, *École de Cavalerie*, 78, 79, 80, 108, 91, 140, 152.

93. De Pluvinel contributed the first detailed prescriptions for achieving airs above the ground through use of pillars, for example, and de la Guérinière the first codification of the principles of the shoulder-in.

94. *An Academy for Grown Horsemen, Containing the Complete Instructions for Walking, Trotting, Cantering, Galloping, Stumbling, and Tumbling* (1787), and *Annals of Horsemanship: Containing Accounts of Accidental Experiments, and Experimental Accidents Both Successful and Unsuccessful* (1791), were both attributed to Geoffrey Gambado and illustrated with caricatures by H. W. Bunbury. "Geoffrey Gambado" evidently was a pseudonym, perhaps of Bunbury. The initial editions of the two works were sometimes rebound as one volume in the 1790s, and appear generally to have been published that way in the early nineteenth century.

95. Thompson, *Rules for Bad Horsemen*, 1765, vi–vii; and 1775, n.p. A reprint in 1793 of the third edition of Thompson's work carried "An Advertisement from the Editor" on the importance of keeping the work available. Three decades later, John Badcock, using the pseudonym John Hinds, published *Rules for Bad Horsemen: Hints to Inexpert Travellers; and Maxims worth remembering by the most Experienced Equestrians. A New Edition, with Modern Additions* (1830).

96. Anon, *Monthly Review,* 244. Inserting the knife a bit deeper, the reviewer concludes: "If there be any such person as J.L Jackson, Esq; we shall only add, that *he is a very modest gentleman*" (Anon, 244). After a good deal of looking, I can find no reference to a J. L. Jackson other than as author of this book.

97. Jackson's *The Art of Riding* is an ornately written treatise of 54 pages. Hughes's *The Compleat Horseman,* a similarly stylized treatise of 58 pages, plagiarizes the first 43 pages of Jackson (or of Thompson): the points and paragraphs, and their sequence, are the same, and the language differs only in word order. O'Reilly's *The Art of Horsemanship,* a shorter treatise of 59 pages (in large font), plagiarizes the first 27 pages of Jackson (or Hughes or Thompson). More direct in style, it trims the rhetorical fat from its predecessors and omits the second half of Jackson's text altogether (since O'Reilly addresses inexperienced horsemen only).

98. Following a passage contrasting equine and human reason, Jackson adds: "But hold! we are got into a strain of philosophizing, and strayed from the subject we were upon; and from laying down rules for good horsemanship, are prating about his capacity and natural endowments, and comparing his abilities with those of the human soul, a digression we should not have fallen into, had not the nature of the subject led us, as it were insensibly, to speak of it" (Jackson, *The Art of Riding,* 51–52).

99. Jackson, preface to *The Art of Riding,* iii, iv.

100. Jackson, *The Art of Riding,* 1, 7, 29, 41.

101. Jackson, *The Art of Riding,* 3, 25. Though Jackson finds some *manège* training useful, he draws the line at *haute école*: airs above the ground are "needless toys" and "a matter of foolish delight rather than of any real use" (Jackson, 48–49).

102. Courtney, "Berenger, Richard," 326. The entry also notes, "Berenger outlived his means, and was obliged for some years to confine himself to his official residence in the King's Mews, then a privileged place against the attacks of bailiffs" (Courtney, 327).

103. Claude Bourgelat (1712–1779) was a pioneer of veterinary medicine, founder of the first veterinary school (in 1762, in Lyons, France), author of many seminal medical works, such as *Art Vétérinaire, ou Médecine des Animaux* (1767), and an avid and accomplished horseman.

104. Berenger, *A New System of Horsemanship,* 2–3.

105. Berenger, *A New System of Horsemanship,* see 10, 54; 3; 12; 32; 64; 48, 65, 132; 113; 10; 91.

106. Berenger, *A New System of Horsemanship,* 20; 118, and see 102; 64; 77–78; 128; 132; 54.

107. Berenger, *The History and Art of Horsemanship,* 1:215. To the twenty-one chapters of his earlier translation of Bourgelat, Berenger here adds a twenty-second chapter on the pirouette.

108. Berenger, *The History and Art of Horsemanship,* 1:1, 3, 5, 164, 168. Berenger notes in passing that "those persons who professed the science of arms were obliged to learn the art of managing their horses, in conformity to certain

rules and principles: and hence came the expression of learning to *ride the great Horse*" (Berenger, 1:170); and he also notes that Robert Dudley, Earl of Leicester, introduced *Claudio Cortio* [*sic*] to the English Court, and that Corte and other Italian horsemen "laid the foundation of the *Manege* in England" (Berenger, 1:183).

109. De la Guérinière, *École de Cavalerie*, 2; Berenger, *The History and Art of Horsemanship*, 1:214.

2. *Manège* to Field

1. Lawrence, *A Philosophical and Practical Treatise*, 120. Lawrence proposed "that the Rights of Beasts be formally acknowledged by the state, and that a law be framed upon that principle, to guard and protect them from acts of flagrant and wanton cruelty, whether committed by their owners or others" (Lawrence, 123). Absent such a law, animals are "property" and, as such, protected from cruelty by others, but not from their owners, whose rights include abusing them at will.

2. DiMarco, *War Horse*, 207–8.

3. De Pluvinel, *Le Maneige Royal*, 89, 93; de la Guérinière, *École de Cavalerie*, 166; Cavendish, *A General System of Horsemanship*, 14.

4. De la Guérinière, *École de Cavalerie*, 191, 192, 100, 190. "Early writers [on dressage and equitation]," Littauer observes in *The Development of Modern Riding*, "claimed . . . that many High School movements were useful in war . . . later ones extolled the practicality of the simpler parts of their method for the cavalry" (Littauer, 83).

5. Cavendish, *A General System of Horsemanship*, 133.

6. Poscharnigg, *Austrian Art of Riding*, 29, vii, 108, 116, 66, see 58–75. Poscharnigg argues, in essence, that a Spanish-Neapolitan tradition of equitation originating in the sixteenth century metamorphosed into an Austrian tradition that established itself in the seventeenth century, achieved its classical moment in the eighteenth century, saved dressage from decadence in the nineteenth century, and entered modernity in the twentieth century—an evolution shaped, in large part, by ongoing innovations in military technology that demanded ongoing modifications to cavalry tactics and training of horses and men.

7. Poscharnigg, *Austrian Art of Riding*, 36, 95, 96, 115.

8. Allen, translator's introduction to Eisenberg, *The Art of Riding a Horse*, 14; Eisenberg, *The Art of Riding a Horse*, 21. Eisenberg reasons that Newcastle and Reganthal, "one of the greatest Masters of our Century" (and Eisenberg's teacher), already had explained dressage both brilliantly and thoroughly. Eisenberg, for his part, would represent their principles and his own with brevity and beauty: "The rider in the illustration represents Monsieur de Reganthal mounted, where I tried more to depict the effect of his seat, than the likeness of his person" (Eisenberg, 102).

9. Eisenberg, *The Art of Riding a Horse*, see 21–32, 64, 54, 38.

10. Eisenberg, *The Art of Riding a Horse*, 130; 22, 24, 28; 128; Allen, translator's introduction to Eisenberg, *The Art of Riding a Horse*, 13; Eisenberg, 54, 130; 19.

11. Pevsner, introduction to Weyrother, *Fragments from the Writings*, 11. When Weyrother's pupils note in their foreword to the German edition of 1836 that his "all too short career . . . did not permit him to complete the work, only a few fragments of which remain to be passed on" (unsigned foreword to Weyrother, *Fragments from the Writings*, 21), they point to a not uncommon problem. While many masters over the centuries codified their theories and practices in systematic treatises, others for various reasons did not produce such work. Masters from Sidney Medows to Federico Caprilli, for example, come to us in more fragmentary and mediated forms, such as compilations of unpublished papers, or student notes from training sessions.

12. Weyrother, *Fragments from the Writings*, 25, 25, 27.

13. Weyrother, *Fragments from the Writings*, 33, 95, 117. Poscharnigg correctly calls Weyrother's reversion to harsh sixteenth-century methods "crass brutality in order to quickly train a cheap cavalry horse" (Poscharnigg, *Austrian Art of Riding*, 108).

14. Foreword by Weyrother's pupils to Weyrother, *Fragments from the Writings*, 21; Weyrother, 103; Poscharnigg, *Austrian Art of Riding*, 108.

15. Poscharnigg, *Austrian Art of Riding*, 95; Chenevix-Trench, *A History of Horsemanship*, 166; and see Littauer, *The Development of Modern Riding*, 89.

16. The brief but influential *Dialogues sur l'équitation* (1835) could be the offspring of a marriage between de Pluvinel and Jonathan Swift—specifically, part 4 of Swift's *Gulliver's Travels* (1726), "A Voyage to the Land of the Houyhnhnms," where Gulliver discovers an enlightened, rational, well-spoken equine race, Houyhnhnms, ruling over a benighted, irrational, loutish humanoid race, Yahoos. Like de Pluvinel's *Le Maneige Royal*, Baucher's *Dialogues* comprises a conversation among three parties and the conversion of one of them; and like *Gulliver's Travels*, Baucher presents one of his three parties, a horse who objects to the horseman's brutal equitation, as the more sympathetic, rational, and moral of the two—at least until the horseman comes to understand "these two expressions which encompass all the principles of equitation: suppling and position," an understanding that he achieves after attending a "clear and precise presentation of [Baucher's] new method" (Baucher, *Dialogues*, 182, 186). The first eleven editions of the *Nouvelle Méthode* were unrevised because of Baucher's publishing contract. Once it expired in 1863, Baucher was free to publish the revised twelfth edition of 1864 and thirteenth edition of 1868. An English translation, based on the Ninth Paris Edition, was published in Philadelphia in 1851 and in Great Britain in 1852 (see note 29 below).

17. Seeger, *M. Baucher and His Art*, 116. For a detailed discussion of Aubert's polemic, *Quelques Observations sur le Système de M. Baucher pour dresser les chevaux* (*Some Observations on the System of M. Baucher for Schooling Horses*) (1842), see Nelson, *Baucher*, 56–68.

18. Oliveira wrote to his disciple Michel Henriquet: "I have invented nothing at all . . . I find that La Guérinière and Baucher have already discovered everything" (quoted in Henriquet, *30 Years with Master Nuno Oliveira*). Decarpentry opens *Baucher et son École* (*Baucher and His School*): "From 340 B.C., when Xenophon wrote the first treatise on equitation that would come down to us, until 1842, when Baucher published his new method, riders of all countries had used nearly identical procedures to educate and direct their horses" (Decarpentry, *Baucher and His School*, 1).

19. Decarpentry published his book-length *Baucher et son École* in 1948, and Jean-Claude Racinet his book-length *Racinet Explains Baucher* in 1997. Hilda Nelson's *François Baucher, The Man and His Method* (1992), her edition of *A New Method of Horsemanship* and *Dialogues on Equitation,* contains an introduction, "François Baucher and the Controversy in Nineteenth-Century French Equitation," comprising eight sections plus an appendix on Baucher's rival Le Comte D'Aure and totaling 105 pages. Nelson's subsequent *Alexis-François L'Hotte, The Quest for Lightness in Equitation* (1997), her edition of General L'Hotte's *Questions équestres* (1895), contains introductory chapters on Baucher, d'Aure, and Baucher and d'Aure together. De Bragança's *Dressage in the French Tradition* (2005) includes a 40-page chapter on "Baucher's System," and Littauer's *The Development of Modern Riding* (1962) includes a 20-page chapter on "Baucher versus d'Aure." Discussions of varying lengths appear in other books and articles.

20. Felton, *Masters of Equitation,* 36.

21. Nelson, *Baucher,* 6. Through the ninth edition of 1850, Baucher's title, *Méthode d'équitation basée sur de nouveaux principes,* was accompanied by a list in much smaller font of supplementary items, including, *Une Théorie sur les means d'obtenir une bonne position du cavalier* (A Theory on Means for Obtaining a Correct Position in the Rider). The English translation, based on that edition, carries an abbreviated title and subtitle: *New Method of Horsemanship, Including the Breaking and Training of Horses, with Instructions for Obtaining a Good Seat.* As Nelson observes, and Baucher acknowledged, he paid more attention to the rider following the early editions.

22. Littauer, *The Development of Modern Riding,* 115; see also: Decarpentry, *Baucher and His School,* 29, 30; Nelson, *Baucher,* 22; and Nelson, *L'Hotte,* 44.

23. De Bragança, *Dressage in the French Tradition,* 89–90; see also: Decarpentry, *Baucher and His School,* 35; Nelson, *Baucher,* 22. The final citation is also from Nelson, *Baucher,* 22.

24. Decarpentry, *Baucher and His School,* 31.

25. De Bragança, *Dressage in the French Tradition,* 95. "We know that the Old School used *general actions,*" de Bragança continues, "and that Baucher rejected them, preferring *partial actions,* because according to him, it was easier to combat resistances one by one. . . . It was after having exercised the different parts of the horse (in the *partial actions*) that Baucher started the *general actions* [via the *effet d'ensemble*]" (de Bragança, 116–17). With respect to Baucher's strong emphasis on

suppling, Nelson writes, "The aim of *assouplissement* [suppling] is to make the horse yield his own force to the horseman's impulsions" (Nelson, *Baucher,* 23).

26. Baucher nicely defines *ramener* as "the perpendicular position of the head, and the lightness that accompanies it," and *rassembler* as "the reunion of forces at the centre of gravity" (Baucher, *Nouvelle Méthode,* 224, 238).

27. De Bragança, *Dressage in the French Tradition,* 92, 103.

28. De Bragança, *Dressage in the French Tradition,* 105; Albert, cited in Nelson, *Baucher,* 57; Nelson, *Baucher,* 100.

29. Decarpentry, *Baucher and His School,* 83; and see de Bragança, *Dressage in the French Tradition,* 97. "Certainly one does not find any trace of the Second Manner in the 11 earlier editions," Decarpentry observes, "because they were nearly a copy of the 1842 text in conformance with clauses in Baucher's contract with his editors, which did not expire until 1863. However, it suffices to go through the works of his students to find, in their chronological order, the modifications that the Master brought to the system in his oral teaching" (Decarpentry, *Baucher and His School,* 83). Commentators speculate that the modifications may have resulted from Baucher's recognition that his method was being misused by novices (Nelson, *Baucher,* 31), or from a serious accident in the circus in 1855, when, as Littauer reports, "the circus chandelier fell on [Baucher]. He escaped death, but his right leg was smashed, and he never rode for an audience again" (Littauer, *The Development of Modern Riding,* 111).

30. See de Bragança, *Dressage in the French Tradition,* 99. Decarpentry writes, "But the *effet d'ensemble,* which, in the First Manner, constitutes the *normal and habitual* means to re-establish lost lightness is not used anymore with this goal in mind." The *effet d'ensemble,* instead, now constitutes "the *ultimate* means of domination" (Decarpentry, *Baucher and His School,* 84). Nelson essentially concurs: "Baucher was not replacing one method with another, that is, *l'effet d'ensemble* by *main sans jambs, jambs sans main,* but was saying that each method has its validity and use, depending upon the experience of the rider and the type of equitation that is being practised" (Nelson, *L'Hotte,* 50).

31. Baucher, *Nouvelle Méthode,* 54, 56.

32. Baucher, *Nouvelle Méthode,* 59.

33. Baucher, *Nouvelle Méthode,* 65, 88.

34. Baucher, *Nouvelle Méthode,* 70, 122, 181, 209.

35. Baucher, *Nouvelle Méthode,* 122, 169–70. Baucher says of the mystifiers that collection "has been a great deal talked about by people, as they have talked about Providence, and all the mysteries that are impenetrable to human perception. . . . The false principles propagated on this subject have made the horse the plaything and the victim of the rider's ignorance" (*Nouvelle Méthode,* 169).

36. Baucher, *Nouvelle Méthode,* 130, 147.

37. Baucher, *Nouvelle Méthode,* 39–40, 41–42, 52.

38. Baucher, *Nouvelle Méthode,* 58, 225, 69, 73. Not surprisingly, Baucher condemns horsemen who "have forged bits of so strange and various forms, real

instruments of torture" (*Nouvelle Méthode,* 105), and advocates the use of simple and mild bits, "for the bit and the hand are as one, and a good hand is the perfection of a rider" (226). Felton writes, "In the last years of [Baucher's] life, he completely changed his methods to the point where . . . he used only the snaffle bridle" (Felton, *Masters of Equitation,* 36).

39. De Bragança, *Dressage in the French Tradition,* 135; Felton, *Masters of Equitation,* 35. Le Comte d'Aure had left an early military career in 1817 to become a pupil-*écuyer* at the École de Versailles. Following his appointment as director in 1827, the École closed in 1830, and d'Aure turned to teaching well-born students in private *manèges.* Returning to military life, he served as *écuyer en chef* of the École de Cavalerie from 1847 through 1854. See Nelson, *L'Hotte,* 52–74, for further details.

40. As Baucher writes in his conclusion to the *Nouvelle Méthode,* "It was the army, above all, that occupied my thoughts. Though counting many skillful horsemen in its ranks, the system which they are made to follow . . . is the true cause of the equestrian inferiority of so many, as well as of their horses being so awkward and badly broken. . . . Yet it is indispensable that a cavalry officer be always master of his horse: the uniformity of manoeuvres, the necessities of command, the perils of the battle-field, all demand it imperatively" (Baucher, *Nouvelle Méthode,* 245, 246–47).

41. Nelson, *L'Hotte,* 42; Felton, *Masters of Equitation,* 36. Baucher's commentators have discussed the role and status of the circus at length. See, for example, Nelson, *Baucher,* 3–4, 15–16 and passim; Nelson, *L'Hotte,* 43–44 and passim, especially chapter 8, "Amazones and *écuyères* of the nineteenth century," 111–34; and Nelson, *The Ecuyere of the Nineteenth Century in the Circus* (2001). See also de Bragança, "Baucherism and the Circus," *Dressage in the French Tradition,* 85–87. The latter sums it up nicely: "True dressage began to appear in the circus at the beginning of the XIX century. . . . It was in the golden age of the circus that Baucher made his appearance" (de Bragança, *Dressage in the French Tradition,* 86). For a twentieth-century text on training horses for the circus, see H. J. Lijsen, *Classical Circus Equitation: Liberty, High School, Quadrilles and Vaulting* (1956; repr. 1993).

42. Baucher was virtually unknown in the 1830s, Nelson writes, while "d'Aure was recognized as an outstanding horseman, a dare-devil riding at break-neck speed, who, at the *École de Versailles,* was already reacting against the formulae and routine of *manège* riding, practicing exterior equitation and the rising trot, and using the English saddle" (Nelson, *Baucher,* 5). But Felton holds that "the publication of [d'Aure's] important book, 'Traité d'Equitation,' in 1834 was brought about . . . by the feeling that he must write something to combat the influence of Baucher whose 'Dictionnaire Raisonné d'Equitation' had been published in the previous year" (Felton, *Masters of Equitation,* 34).

43. The letter of rejection from the Comité to Baucher indicated that his method allowed insufficient time for training men, was "more in keeping with *manège* equitation, rather than the training of military horses," and was detrimental to military horses (quoted in Nelson, *Baucher,* 49). The unspoken reasons included

Baucher's social class and civilian status. Ironically, Baucher's method was designed to train problem horses, plentiful in the army, and to do it quickly, a military priority.

44. Littauer, *The Development of Modern Riding*, 118; and see Felton, *Masters of Equitation*, 34. For private purposes, d'Aure practiced *haute école* equitation throughout his lifetime. A true believer, he criticized Baucherism, in large part, because it led the public to confuse "pure Classical equitation [as it had been practiced at the École de Versailles] and the equitation that was currently being presented in the circus under the guise of *haute école* equitation" (Nelson, *Baucher*, 51).

45. Littauer, *The Development of Modern Riding*, 119; Nelson, *L'Hotte*, 57, 71.

46. See Nelson, *L'Hotte*, 75–78; Nelson, *Baucher*, 68, 21; Littauer, *The Development of Modern Riding*, 123.

47. The French were not alone in equestrian Anglomania. As Poscharnigg observes, thoroughbreds "became the horses for the [European] equestrian élite who knew how to handle them," though they were "dangerous for stupid riders." Concurrently, "campaign equitation and English steeplechase riding" became more popular than high school riding for training military horses (Poscharnigg, *Austrian Art of Riding*, 93, 98). In *Cavalry in Future Wars* (1899, 1906, 1909), General Friedrich von Bernhardi would deplore the unfavorable influence of "Anglomaniacs and faddists" on German cavalry (see Bernhardi, *CFW*, 184–85).

48. Huggins, *Horse Racing and British Society*, 203; 210; 209 and 284.

49. De Bragança, *Dressage in the French Tradition*, 82. See Littauer, *The Development of Modern Riding*, 70; Nelson, *Baucher*, 3–4; Nelson, *L'Hotte*, 55. Poscharnigg, *Austrian Art of Riding*, 94.

50. "I certainly do not mean to say," de Bragança writes, "that liberalism and romanticism had as doctrines, an influence on horsemanship, rather that the romantic culture and the liberal social climate with its revolutionary ideals created an innovative climate into which Baucher found himself integrated. This social atmosphere encouraged him to give form to a revolutionary form of dressage" (de Bragança, *Dressage in the French Tradition*, 83–84). And see Nelson, *Baucher*, 2.

51. "In an effort to prove himself an absolute innovator," Tomassini writes, "Baucher flatly refused the principles of what classical riding was up until then, especially the fundamental doctrine of La Guérinière" (Tomassini, *The Italian Tradition of Equestrian Art*, 247).

52. The cult of Baucher as rebel and virtuoso may explain in part why the Romantic writer Theophile Gautier and painter Eugène Delacroix were ardent Baucherists. As a social phenomenon, the Baucher cult closely resembled the cults of Paganini and Liszt, as well as, save the element of doomed genius, those of Byron and Shelley.

53. Nelson, *Baucher*, 5. Nelson stands on firmer ground when she writes, "Baucher's method, while termed nouvelle (new), remains, nevertheless, in the idiom of what is known as *manège, savante*, academic or Classical equitation" (Nelson, *Baucher*, 5). Baucher, on this view, revolutionized a system from within, rather than rebelled against the system.

54. As Tomassini writes, "The eighteenth is still a century of French hegemony, although in the equestrian field begins to manifest the Franco-German dualism, which will result soon in open rivalry between the two cavalries" (Tomassini, *The Italian Tradition of Equestrian Art,* 229).

55. Seeger, *M. Baucher and His Art,* 115–17. In Seeger's view, Baucher's rejection of the trot as the primary training gait, whatever its equestrian rationale, was one form of his trickery: "It is with reason that M. Baucher sharply cuts short his work at the trot, first because it displays the faults in his system too obviously and secondly because his work at the canter permits him to fool the public more easily" (Seeger, 119).

56. De Bragança, *Dressage in the French Tradition,* 133; Seeger, *M. Baucher and His Art,* 116, 123, 127, 131, 134, 136, 129, 139.

57. Seeger, *M. Baucher and His Art,* 130, 130, 132, 132, 117, 158, 159.

58. Steinbrecht had left Plinzner with a book still in fragments. "Since he had given me unlimited power," Plinzner notes in his preface to the first edition, "I had the choice of either publishing the fragments or independently completing them . . . into a complete whole." He chose the latter course, adding that the chapters on the canter, piaffe, passage, and the airs above the ground, as well as the epilogue, "have come exclusively from my pen" (Plinzner, preface to Steinbrecht, *The Gymnasium of the Horse,* xii–xiii). Plinzner, in turn, charged Heydebreck to carry Steinbrecht's vision forward, so Heydebreck introduced the *Gymnasium's* fourth edition in 1935, adding commentary on Plinzner's deviation from Steinbrecht's teachings.

59. Steinkraus, foreword to Steinbrecht, *The Gymnasium of the Horse,* vi; Pevsner, introduction to Weyrother, *Fragments from the Writings,* 11.

60. Plinzner, preface to Steinbrecht, *The Gymnasium of the Horse,* xii.

61. Steinbrecht, *The Gymnasium of the Horse,* 1, 19, 145–46, 56, 59, 292.

62. Steinbrecht, *The Gymnasium of the Horse,* 71 and 199, 73 and 79, 49, 132, 85.

63. Steinbrecht, *The Gymnasium of the Horse,* 7, 43, 44, 48, 64, 125, 127.

64. Steinbrecht, *The Gymnasium of the Horse,* 311. Steinbrecht's disgust for "pedantry" and "false erudition" (Steinbrecht, 33, 87), for the degradation of equestrian art to "philistinism and puppetry" (125), follows from his commitment to "the principle that all movements of true, classical dressage have direct or indirect practical utility" (232).

65. Williams, preface to *H.Dv.12: Army Riding Regulations,* ix; Hess, foreword to *H.Dv.12: Army Riding Regulations,* xiii.

66. De Bragança, *Dressage in the French Tradition,* 132. Felton makes the same point, if more glibly: "The French willingly sacrifice complete control and subjection of the horse in order to obtain the airy grace of a ballet dancer. The Germans sacrifice the freedom and grace of the French-schooled horse to achieve the precision and the technically exact performance of a drill team or a dancer in a Rockette chorus" (Felton, *Masters of Equitation,* 40).

67. Fillis, *Breaking and Riding,* 342; echoed by Felton, *Masters of Equitation,* 38. Chenevix-Trench writes, "Just as d'Aure had aimed at combining the good points of German dressage and English cross-country riding, and at Saumur in the 1890s the teaching was a synthesis of d'Aure and Baucher, so Fillis tried to adapt to cross-country riding the principles of the High School" (Chenevix-Trench, *A History of Horsemanship,* 170). Felton puts it more cautiously: "Although . . . Fillis was interested primarily in school riding, he did also teach to a slight extent at least outdoor riding . . . and he even had some comments to offer on jumping" (Felton, *Masters of Equitation,* 37). Fillis's contemporary, the military historian Lieut.-Col. F. N. Maude, wrote in 1903, "Seeing where the strong and the weak points of the school riding and the steeplechase system lay, [Fillis] set himself to combine the two into one working method, and judging both by his book and his practice, he succeeded most admirably" (Maude, *CPF,* 235).

68. The text of *Questions équestres,* translated by Hilda Nelson, is the center-piece of Nelson's *Alexis-François L'Hotte, The Quest for Lightness in Equitation* (1997), a work as indispensable as her earlier *François Baucher, The Man and His Method.* The introductory materials to *L'Hotte,* totaling 145 pages, include a biography of L'Hotte, studies of Baucher, d'Aure, and Rousselet in relation to one another and, especially, to L'Hotte, and a detailed summary and analysis of *Questions équestres.*

69. L'Hotte, *Questions équestres,* 194. The anonymous editor of a posthumous reissue of *Questions équestres* (1906) prefaced the work as L'Hotte's distillation of notes taken over "sixty years of practice and study" into a concise body of princi-ples, goals, and procedures presenting "a simple and clear method" of equestrian art and representing "a veritable philosophy of equitation" (quoted in Nelson, *L'Hotte,* 149).

70. L'Hotte, *Questions équestres,* 151, 201, 202. Nelson points out that L'Hotte schooled military horses using "only the equitation of d'Aure," but his personal horses using both the outdoor precepts of d'Aure and the *manège* pre-cepts of Baucher (Nelson, *L'Hotte,* 80). She astutely adds, "When [L'Hotte] dis-cusses and defends d'Aure [in *Un Officier de cavalerie*] he does so with greater enthusiasm [than with Baucher]. While he defends Baucher and calls him 'the greatest *écuyer who ever lived,*' there seems to be missing . . . the same degree of vigour" (Nelson, *L'Hotte,* 140–41).

71. L'Hotte, *Questions équestres,* 153, 167, 154, 155, 157–58, 159. "As a gen-eral rule and in view of his quest for lightness," L'Hotte later adds, "the experienced *écuyer* knows that he must not seek the *rassembler* before he has introduced the *ramener.* . . . The *ramener,* as it is understood in high equitation, has little to do with the position of the horse's head. It lies, first of all, in the submission of the jaw which is the first joint that receives the effect of the hand. . . . The *ramener* . . . is less an unchanging position of the head, but, rather, a general condition of the submis-sion and pliancy of all the joints and muscles" (L'Hotte, *Questions équestres,* 196).

72. L'Hotte, *Questions équestres,* 159, 178, 180, 183, 199, 156. In "striking off at the canter," for example, "one must wait for the horse, for it is he who executes

the movement" (L'Hotte, 176). Likewise jumping: "[The horse's] instinct alone, rather than anything the rider can do, will guide him in using the different resources that nature has given him" (184). Tact, put differently, also entails patience and trust.

73. "The characteristics of Classical or *savante* equitation are the opposite of those that contribute to the success of circus equitation," L'Hotte writes. "With Classical equitation, the rider must always remain correct, if not flawless. . . . The horse must obey at the lightest touch of the aids which must always be discreet. . . . It is nature that this equitation takes as guide and not the extraordinary or the eccentric that is sought" (L'Hotte, *Questions équestres,* 195). This may be a left-handed compliment to Baucher as a true *manège* rider who also could perform circus stunts.

74. L'Hotte, *Questions équestres,* 157, 162, 190, 192 and 200.

75. Felton goes so far as to characterize Clemenceau as Fillis's "ghost writer" (Felton, *Masters of Equitation,* 38). Hayes was the author of several notable works, including *Riding on the Flat and Across Country* (1881) and *Points of the Horse,* published in seven editions between 1893 and 1969. Alice M. Hayes, his spouse, was the author of the influential treatise, *The Horsewoman: A Practical Guide to Side-Saddle Riding* (1903).

76. Fillis, preface to *Breaking and Riding,* vii–viii.

77. Fillis, *Breaking and Riding,* 5–6, 5, 184, 174–75, 177. Fillis notes that he is not contradicting his initial statement, "I break in only thorough-breds for my own use" (Fillis, 1), when he later writes, "The half-bred is the best animal for war." The thoroughbred lacks requisite endurance, and, in any case, "To make use of a Thoroughbred, one must know more than ordinary cavalrymen do about riding" (213, 214n).

78. Fillis, *Breaking and Riding,* 83, 299. "A broken horse," Fillis writes, "is not a machine which requires only to be wound up, but is a living creature . . . who requires to be constantly kept in the discipline of work" (Fillis, 259n). And that discipline, he repeats often, depends entirely on "appropriate punishment and reward" (42) to ensure submission and obedience.

79. Fillis, *Breaking and Riding,* 314, 23, 129, 130n, 2, 64, 70, 104, 243.

80. Fillis, *Breaking and Riding,* 332, 332n, 63, 56. Fillis's legerdemain in his defense of Baucher continues with a specious explanation for why Baucher never rode outside: "Baucher being a reformer and consequently a seeker, had no pleasure in leaving a horse to himself, as is done when hacking. He devoted all his life to his work in order to show us the way, which was the only thing that interested him" (Fillis, 343).

81. Fillis, *Breaking and Riding,* 213–14, 225–26; Felton, *Masters of Equitation,* 39; Chenevix-Trench, *A History of Horsemanship,* 171.

82. Beudant, *Horse Training: Out-door and High School,* vii.

83. Faverot de Kerbrech, *Methodical Dressage,* 2.

84. Faverot de Kerbrech, *Methodical Dressage,* ix. On Baucher's riding accident, see note 29 above.

85. Faverot de Kerbrech, *Methodical Dressage*, 3, 113, see 97, 82, 95.

86. Faverot de Kerbrech, *Methodical Dressage*, 117, 20, 69, 110, 119.

87. Faverot de Kerbrech, *Methodical Dressage*, 118, 119.

88. Beudant, *Horse Training: Out-door and High School*, 25.

89. Beudant, *Horse Training: Out-door and High School*, 9–10, 16, 22, 23, 22.

90. Beudant, *Horse Training: Out-door and High School*, 4, 21, 4, 33, 101.

91. Beudant, *Horse Training: Out-door and High School*, 3, 11, 16, 108–9.

92. Beudant, *Horse Training: Out-door and High School*, 18, 60–61, 100.

93. Beudant, *Horse Training: Out-door and High School*, letter "To General of Division Juinot-Gambetta," vii; 3; 105.

94. Beudant himself observed, "The routine of a rigorous military theory and the faint-heartedness of the training given our horses makes them all too often completely impotent when facing the difficulties and dangerous obstacles habitual with our foreign comrades, notably the Italians" (*Horse Training: Out-door and High School*, 120).

95. See Caramello, "Revisiting Piero Santini, Apostle of Forward Riding."

3. Light-Horse, Dragoons, and Others

1. DiMarco, *War Horse*, 193; Badsey, *Doctrine and Reform in the British Cavalry*, 1. As a way of indicating the relative scale of the cavalries fielded by the Central Powers and Allies in August 1914—respectively, 100,000 and 150,00—Kenyon notes in *Horsemen in No Man's Land* that "the cavalry portion of Napoleon's *Grande Armée* of 1812, possibly one of the largest single military forces ever previously assembled . . . amounted to only 80,000 horsemen"—hardly a trivial number (Kenyon, 17).

2. The books under discussion in this chapter were published roughly between 1750 and 1950, with the preponderance between roughly 1880 and 1920. They include learned treatises on cavalry both broad and narrow in scope; general introductions to the arm, with surveys of its components, frequently addressed to young officers; and contemporary histories for military and civilian audiences. They do not include official regulations, technical manuals, or, with a few exceptions, biographies, autobiographies, and memoirs published in the period. Since several writers are represented by two or more books, abbreviations of titles are used in citations: Bernhardi, *Cavalry in Future Wars* (*CFW*), *Cavalry in War and Peace* (*CWP*), *Germany and the Next War* (*GNW*), *How Germany Makes War* (*HGMW*), *The War of the Future* (*WF*); Carter, *Horses, Saddles, and Bridles* (*HSB*), *The American Army* (*AA*); Childers, *War and the Arme Blanche* (*WAB*), *German Influence on British Cavalry* (*GIBC*); Denison, *Modern Cavalry* (*MC*), *A History of Cavalry from the Earliest Times* (*HC*); Liddell Hart, "After Cavalry—What?" ("*ACW*"), *The Remaking of Modern Armies* (*RMA*); Maude, *Cavalry Versus Infantry* (*CVI*), *Cavalry: Its Past and Future* (*CPF*); Nolan, *The Training of Cavalry Remount*

Horses (*TCRH*), *Cavalry: Its History and Tactics* (*CHT*); and Wagner, *The Service of Security and Information* (*SSI*), *Organization and Tactics* (*OT*).

3. Badsey, *Doctrine and Reform in the British Cavalry,* 3; Wagner, *OT,* 1, and see Jones, *From Boer War to Great War,* 11–12; Bernhardi, *HGMW,* 19. Bismarck clarified the relationships among tactics, evolutions, and maneuvers in 1818: "Tactics are the art of placing troops in positions for battle, and of manoeuvring them with advantage: simple positions and movements of troops are called evolutions, the combination of evolutions is called a manoeuvre; and the art of applying these manoeuvres to the operations of war, in such a manner as to attain the object in view, is called tactics" (Bismarck, *Lectures on the Tactics of Cavalry,* 8).

4. See, for example, Goldman, *With General French and the Cavalry in South Africa,* 408.

5. Bismarck, *Lectures on the Tactics of Cavalry,* 301–2; Nolan, *CHT,* 39; Wagner, *OT,* 57, 238.

6. Childers, *WAB,* 21; Wagner, *OT,* 63; Rimington, *Our Cavalry,* 101. Childers's comment can serve to illustrate the scores of comments made about mobility as an "essential condition of the success of cavalry" (McClellan, *European Cavalry,* 164), or as the "prime condition of efficiency in its strategical activity" (Bernhardi, *CFW,* 169), or of the critical importance of "securing under *all conditions* the maximum possible degree of mobility" (Maude, *CPF,* 274), or of mobility as the cavalry's "chief strength" (Goldman, *With General French and the Cavalry in South Africa,* ix), or about the application of mobility to any number of specific tactical situations. Whenever the efficacy or future of cavalry came into question, writers invariably invoked the arm's mobility, such as in the 1920s, when Malcolm Wheeler-Nicholson argued, "The chief value of cavalry is its value as a highly mobile battle arm" (Wheeler-Nicholson, *Modern Cavalry,* 8). And if mobility is the fundamental value, then the halt is its devaluation, or, as Captain Loir put it in 1912, "A halt . . . is always a period of danger to cavalry. Cavalry can only exist *by* its mobility and *for the sake* of mobility. A halt takes away its only means of action" (Loir, *Cavalry,* 75).

7. Badsey, *Doctrine and Reform in the British Cavalry,* 21; Monsenergue, *Cavalry Tactical Schemes,* xv; see Wheeler-Nicholson, *Modern Cavalry,* 1. Writing in 1881, for example, Carl von Schmidt found "no occasion whatever to fear that [dismounted service] will impair the true cavalry spirit; indeed, it can only gain by it, as our arm will be able to accomplish its object in all situations" (Schmidt, *Instructions for the Training, Employment, and Leading of Cavalry,* 121).

8. Brereton, *The Horse in War,* 68; Hinde, *The Discipline of the Light-Horse,* 57; Bismarck, *Lectures on the Tactics of Cavalry,* 72; Nolan, *CHT,* 144. As Badsey explains, "The power of a mounted charge . . . came from a combination of the speed of the horses and the skill of their riders in maintaining a straight line, so that they all struck the enemy at once: about three-quarters of a ton of horse and rider moving at a gallop of over seven yards a second (440 yards a minute in British regulations)" (Badsey, *Doctrine and Reform in the British Cavalry,* 15). Though constant in a given action, and variable from action to action, intervals between horses were

always close. As Alonzo Gray wrote in 1910, "The more solid the mass at the instant of impact, the greater will be the effect of the shock. The charge should, therefore, be made boot-to-boot" (Gray, *Cavalry Tactics,* 45). Charges, however, had not always been at speed. As Wagner wrote in 1895, "The action of the cavalry [in the Thirty Years War, in the seventeenth century] was essentially by shock, though the charge was still made at a trot." It was in the eighteenth century that Charles XII of Sweden "taught the cavalry to charge at full speed. The true rôle of cavalry was now beginning to be understood" (Wagner, *OT,* 193, 194). The concept of mass plus momentum also pertained to cavalry charges against cavalry. Bernhardi wrote in 1910, "In the mounted combat against cavalry, every effort must be directed towards falling upon the enemy at full gallop in a serried mass, and thus to overthrow him" (Bernhardi, *CWP,* 119); and Rimington in 1912, "In the mounted attack of cavalry on cavalry that side will win which makes use of a wall of mounted men, advancing knee to knee with no intervals showing" (Rimington, *Our Cavalry,* 37).

9. Bernhardi, *HGMW,* 96; Rimington, *Our Cavalry,* 87.

10. Bismarck, *Lectures on the Tactics of Cavalry,* 299; Nolan, *CHT,* 82; Baker, *The British Cavalry,* 43; Coulston, introduction to Nolan, *CHT,* xvii; Neville, *A Treatise on the Discipline of Light Cavalry,* 30; Denison, *MC,* 30; Wagner, *OT,* 218; Schmidt, *Instructions for the Training, Employment, and Leading of Cavalry,* 14. An infantry weapon, the bayonet made a brief, belated, and evidently inconsequential appearance in the conversation on cavalry *arme blanche* by way of "mounted infantry," or "mounted rifles" (see the final sentence of Childers, *GIBC,* 215).

11. On slashing bridle hands in mounted duels, see, for example, Nolan, *CHT,* 47. Janet Macdonald speculates that, in the eighteenth century, "blows aimed at [the] right [sword] forearm, wrist and hand . . . constituted some 70 per cent of cavalry wounds, with another 20 per cent being to the head" (Macdonald, *Horses in the British Army,* 125). For views on cutting and thrusting from 1781 through 1915, see, for example, Warnery, *Remarks on Cavalry,* 16; Bismarck, *Lectures on the Tactics of Cavalry,* 139; Baker, *The British Cavalry,* 59; McClellan, *European Cavalry,* 25, 207; Gray, *Cavalry Tactics,* 26; Hayne, *Lectures on Cavalry,* 57; and see note 99 below. The citation on moral effect is from Baker, 59; and see Gray, 13. For examples of the many manuals on mounted swordsmanship, often printed in multiple editions, see Fawcett, *Rules and Regulations for the Sword Exercise of the Cavalry* (1796) and *Six Engravings, Representing the Six Cuts in the Sword Exercise of the Cavalry* (1803); [Calvert], *Regulations and Instructions for Cavalry Sword Exercise* (1819); and, in the American cavalry, Hershberger, *A Sabre Exercise for Mounted and Dismounted Service* (1844).

12. DiMarco, *War Horse,* 257. Moreover, as Baker advised, if a mounted trooper carried his carbine slung across his back, "the man's back is, to a great extent, protected from a sword cut" (Baker, *The British Cavalry,* 43), a point he may have derived from Nolan (*CHT,* 78).

13. Writers shared consensus but not unanimity on use of the carbine while mounted. Baker considered the practice "difficult and inaccurate," whether at a

gallop or halt, "as the breathing of the horse interferes so materially with the aim" (Baker, *The British Cavalry*, 60); Denison and Schmidt concurred on the proscription. Wagner, however, argued, "Mounted fire action is not frequently used, but it is nevertheless of sufficient value to be seriously considered, and there is nothing to justify the assertion of some European writers [specifically Schmidt] that the trooper's carbine should never be fired from the saddle except as a signal" (Wagner, *OT*, 55). Mobility aside, writers agreed that the infantry rifle outperformed the cavalry carbine in range and accuracy (see Ingelfingen, *Letters on Cavalry*, 238, and Rankin, *A Subaltern's Letters to His Wife*, 180). Jones writes that Boer rifles for example, exceeded British carbines in range by 2,500–3,000 yards to 1,200 yards (Jones, *From Boer War to Great War*, 172).

14. Cavalry in the United States, particularly Confederate cavalry, pioneered the use of the pistol and the tactical "pistol charge," though British and European cavalry, whether through traditionalism or battle conditions, adopted neither. As early as 1868, however, the Canadian officer George T. Denison raised the question, "arisen since the last war in America, as to whether the revolving pistol has not taken the precedence in the *mêlée* over the sword" (Denison, *MC*, 30). He even praised the revolver as a defensive as well as offensive weapon in a melee: "In the American war it was constantly used to ward a sabre cut or parry a thrust, as well as to pour in sharp and deadly discharges" (Denison, *HC*, 429). On machine guns, see notes 109, 110, 111 below.

15. Denison, *HC*, 442; see also Ingelfingen, *Letters on Cavalry*, 128; Bernhardi, *CFW*, 59, 63; and Wheeler-Nicholson, *Modern Cavalry*, 14. Histories of cavalry throughout the period generally subscribed to the "great man" orthodoxy, and thus showed more interest in generals than in their troops, often describing the former as "great artists." The Crimean War, however, spawned a heterodoxy that gained purchase over the next half-century and into the Great War: noble troopers and junior officers sacrificed by incompetent, amateurish, aristocratic generals.

16. The epigraph to the 1827 English translation of Bismarck's *Lectures on the Tactics of Cavalry* observes, "Reading and Discourse are requisite to make a Souldier perfect in the Art Military, how great soever his practical knowledge may be" (attributed to *Observations upon Military & Political Affairs, by General Monck, Duke of Albemarle*, 1671). Colonel F. Chenevix-Trench wrote, appositely, in 1884, "As a matter of fact, the Frederics and Napoleons of history have been thorough students as well as practical soldiers, and an officer who does not read can never, in these days, hope to excel" (Chenevix-Trench, *Cavalry in Modern War*, v). General Douglas Haig cited Moltke's pertinent criticism of the Prussian cavalry of 1866: "Thus, approved methods were altogether forgotten, a heavy indictment against the manner in which military history has been studied during the years which have elapsed since the Napoleonic wars" (Moltke, quoted in Haig, *Cavalry Studies*, 11). With respect to recent wars, cavalry writers in the late nineteenth and early twentieth centuries referred for lessons most often to the American Civil War, the Franco-Prussian War, and, after the turn of the century, the Boer War.

British and German writers—such as Denison, Chenevix-Trench, Goldman, French, Haig, Childers, Ingelfingen, and Bernhardi—invoked the American Civil War, a touchstone for modern cavalry tactics, but their armies did not adopt its lessons.

17. Maude, *CPT,* 3; Denison, *MC,* 99.

18. Bernhardi, *CFW,* 184; Nolan, *CHT,* 60, 107.

19. For this reason, Warnery advised in 1781, "Soldiers for the cavalry [should be] chosen from a grass country, because the men and horses, which both grow there, are brought up as it were together" (Warnery, *Remarks on Cavalry,* 25); and Bismarck in 1818, "It is of importance, in selecting recruits for cavalry, to choose those only, who, either volunteer, or who have been accustomed to horses from their youth, such as the sons of farmers" (Bismarck, *Lectures on the Tactics of Cavalry,* 144). Increased industrialism and urbanism in the nineteenth century, combined with mass cavalries, made such selectivity less viable.

20. As Badsey summarizes, "In addition to learning the basics of soldiering, cavalry troopers also had to learn to ride and to care for their horses, the horses had to be trained, and together they had to master the various skills and tactical evolutions required for mounted combat. . . . [They] had to be taught not just horsemanship but 'horsemastership' (or sometimes 'horse management'), a mixture of basic veterinary science and country wisdom aimed at keeping horses in good condition. An important part of horsemastership was the care that came from an emotional bond that was encouraged between mounted soldiers and their horses, particularly on active service" (Badsey, *Doctrine and Reform in the British Cavalry,* 10).

21. Denison, *HC,* 418; Rimington, *Our Cavalry,* 203. Training manuals and guides typically advised officers to observe the cardinal adage of horse care, "no foot, no horse," by monitoring the state of their horses' feet and shoes, and by ensuring that farriers followed shoeing regulations. Hinde, in 1778, for example, specified punishment for "Any Farrier who presumes to make shoes after any other than the regimental pattern" (Hinde, *The Discipline of the Light-Horse,* 425); Bismarck, in 1818, wrote, "A horse neglected in shoeing . . . is generally unserviceable for the whole campaign" (Bismarck, *Lectures on the Tactics of Cavalry,* 50); Denison, in 1877, "Every effort should be made by careful feeding, shoeing, saddling, &c., to preserve [horses] in good condition (Denison, *HC,* 437); Gray, in 1910, "The want of horseshoes will render cavalry temporarily unserviceable" (Gray, *Cavalry Tactics,* 164); Loir, in 1912, "A horse with a shoe off cannot march" (Loir, *Cavalry,* 42); and Parker, in 1917, "All officers must understand the principles of proper shoeing and be able to supervise the work of the horseshoers. A trooper should know how to put on a shoe in an emergency" (Parker, *An Officer's Notes,* 160).

22. Bismarck, *Lectures on the Tactics of Cavalry,* 61; Nolan, *CHT,* 189. Improvements in infantry rifles made battle more lethal to horse cavalry, but the degree can be debated. A galloping horse and rider present a large target, to be sure, but also a fast-moving one not so easy to hit. Childers awarded the advantage to the rifleman, given "the growth in the destructive efficacy of the firearm, directed against so large

a target as presented by rider and animal combined" (Childers, *WAB,* 24); but Major LeRoy Eltinge, noting the difference between "target practice" and "battle conditions," observed that in the latter, "A swiftly moving horse is a poor target" (Eltinge, "Notes on Cavalry," 46). Technical innovations, moreover, could gain higher accuracy, but often at the expense of caliber and stopping power. As Maude wrote, "It does matter very much indeed if each bullet which finds its billet drops horse or man dead in its tracks, or whether they can gallop on with half a dozen wounds for a mile or two before they fall to the ground" (Maude, *CPF,* 182; and see Haig, *Cavalry Studies,* 9; and Gray, *Cavalry Tactics,* 40).

23. Andrew Roberts uses this astute phrase in his biography, *Napoleon: A Life* (2014). David's later portrait, *The Emperor Napoleon in His Study at the Tuileries* (1812), provides the counterpoint to the earlier equestrian portrait. Static in composition though not in narrative, it depicts an older Napoleon standing in his study at dawn, having worked through the night on the *Code Napoléon* and about to review his troops: Napoleon, in short, is represented as monarch, tireless civil administrator, and military leader.

24. Pembroke, dedication, *Military Equitation,* n.p. Pembroke is warning the king, diplomatically, that the "wretched system" ill serves His Majesty's political interests. Even more slyly, Pembroke recalls to the king that they "had frequent occasion to *lament together*" the current state of military horsemanship (my italics) and that "You have often done me [the honour] of talking to me upon Horsemanship." Thus, Pembroke uses the dedication, one, to solicit the king for advocacy and resources for the methods to be described, and two, to make clear to others that the king and he are in accord explicitly on the problem and implicitly on its solution. Pembroke adds in his dedication that *Military Equitation,* or "this little work," represents "the outlines only of a more extensive, general one, which I intend to make public hereafter, should I find time to finish it." Obviously, he never found the time. In *An Analysis of Horsemanship,* John Adams succinctly describes *Military Equitation* as a "small tract . . . intended solely for the use of the army. It contains only a few general rules, absolutely necessary for the discipline of the cavalry" (Adams, *An Analysis of Horsemanship,* 1799, 1:ix).

25. Pembroke, *Military Equitation,* 87, and see 76, 97, 99; 21; 19; 83.

26. Pembroke, *Military Equitation,* 8; 23, and see 72; Littauer, *The Development of Modern Riding,* 91. When Littauer writes that "Pembroke's is a very small and unimportant book" in comparison to the earlier eighteenth-century masterworks on dressage by Newcastle, de la Guérinière, and company, he was not making an invidious comparison to dismiss Pembroke, but, rather, was praising him for successfully repurposing French *manège* training for the British military work at hand (Littauer, *The Development of Modern Riding,* 90–91). Pembroke advises, for example, that field maneuvers require precise use of shoulder-in, haunches-in, and rein-back "in all openings and closings of files" (Pembroke, *Military Equitation,* 56), and that close combat requires agile one-handed riding, since the right hand "carries the sword, which is a sufficient business for it" (Pembroke, 16, and see 20,

24). Pembroke also opposes "the silly custom of using strong and heavy" bits (27); details what he considers proper and improper use of work in hand (42–52); and cites Bourgelat, in Berenger's translation, on the trot (61) and Newcastle on resistances (92).

27. Tyndale, *Instructions for Young Dragoon Officers*, 15; *A Treatise on Military Equitation*, 1–2, 3. In contrast to the four editions of Pembroke's *Military Equitation*, Tyndale's *A Treatise on Military Equitation* was published by subscription in one edition only. The work lists five principal subscribers, beginning with His R. H. the Prince of Wales, followed by 129 officers from 21 regiments, followed by four nonmilitary subscribers.

28. Tyndale, *A Treatise on Military Equitation*, 8, 60. Tyndale also takes the sensible if then contrarian tack, "The method I have recommended [in essence, teaching and learning fundamentals of classical equitation and dressage], so far from spoiling a hunting seat, will, in my opinion, give it more force, and be more favourable to the horses; it will also make them go up to, and take their leaps in a much better stile" (Tyndale, *A Treatise on Military Equitation*, 60).

29. Adams, *An Analysis of Horsemanship*, 1799, 1:ii, vii, xv, xix, xxxiv. *An Analysis of Horsemanship* appeared as a double-decker in 1799 and as an expanded triple-decker in 1805, both carrying the subtitle, *Teaching the Whole Art of Riding, in the Manege, Military, Hunting, Racing, and Travelling System.* I can find no biographical information on John Adams, so presume that the name may be pseudonymous; this John Adams, in any case, was not the presidential John Adams.

30. Adams, *An Analysis of Horsemanship*, 1799, 1:182, 188–90, 195. Subsequently, Adams identified and discussed "the difference [between] the three principal seats—the Manege, Military, and Hunting Seats," describing the military seat, a hybrid of the other two, as "still keeping the body upright, but sinking as low as you can without stooping, or projecting the knee before the toe" (Adams, 1805, 2:17). Littauer credits Adams with being, "I believe, the first one to present the technique of jumping at any length (12 pages). Thus he was more advanced in his thinking than either the Earl of Pembroke or Count Drummond de Melfort, who changed nothing but merely abridged the program of the old school." Littauer, a disciple of Caprilli's "Italian method," or "forward seat," adds, "Adams' description of the hunting seat contains amazingly modern ideas. . . . He can be considered a precursor of Caprilli, who developed the Forward Seat one hundred years later. . . . But however vague his description may be, it is the first appearance in print of some of the basic ideas of the Forward Seat." Littauer's one reservation: "The seat Adams intended for the gallop only. He did not go so far as to work it out for the jump," as Caprilli would do (Littauer, *The Development of Modern Riding*, 96–98). Felton likewise credits Adams with advocating "a more balanced seat, not unlike the modern hunting seat" (Felton, *Masters of Equitation*, 49).

31. Adams, *An Analysis of Horsemanship*, 1805, 1:188, and see 229, 253; 190; 204–6; 225. The trope of the army, or one of its divisions, as a machine dates at least to *Military Instructions from the late King of Prussia to His Generals* (1762), an

English compilation of instructions given by Frederick II in the 1740s, with its depiction of an army as an "artificial machine" whose smooth functioning required the vigilant attention of generals (Frederick, *Military Instructions,* 5). Ideally suited to the industrial age of the eighteenth and nineteenth centuries, the inevitable and ubiquitous trope of army as machine was employed as a positive image, for example, by Warnery in 1781 (Warnery, *Remarks on Cavalry,* 22), Bismarck in 1818 (Bismarck, *Lectures on the Tactics of Cavalry,* 52), Baker in 1858 (Baker, *The British Cavalry,* 63, 104), and Rimington in 1912 (Rimington, *Our Cavalry,* 98). Concurrently, however, Bismarck and Nolan each pointed out that the army *might* function like a machine were it not for the *actual* conditions of warfare (Bismarck, 88; Nolan, *CHT,* 33, 97); and Childers praised "the revolution [in musketry] whose essence was the substitution of individual skill and intelligence for those formal, machine-like movements of massed bodies . . . best exemplified in . . . shock action" (Childers, *WAB,* 15). Following the unprecedented destruction visited by and upon mass, mechanized armies in the Great War, the trope of army as productive machine transformed into one of war as destructive machine. See chapter 5, note 9.

32. Freeman's teacher and subsequent partner in a riding-house, and one of many distinguished horsemen over the centuries who rode brilliantly but wrote nothing about it, Sidney Medows (c. 1699–1792) left a legacy of principles that Freeman (1754–1821) intended to consolidate posthumously and share with the military and riding publics. Originally a follower of Newcastle, as Freeman notes, Medows had evolved into an independent and original equestrian thinker whose body of principles derived less from Newcastle than from his own experience and experiments. He was eulogized as "perhaps, the most complete rider of managed horses in the kingdom; and so fond of it was he to the last . . . that, not many hours before his death, he made his servants set him on horseback" (Urban, *The Gentleman's Magazine,* 1236). Berenger dedicated his translation of Bourgelat's *Nouveau Newcastle* to Medows, and, in *The History and Art of Horsemanship,* placed Medows in Newcastle's company, but also noted, "He never yet has thought proper to convey his knowledge to others by means of the *Press,* but . . . *does* more than other people *write.* His *Horse* is his *Pen*" (Berenger, *The History and Art of Horsemanship,* 1:213).

33. Freeman, *The Art of Horsemanship,* i, 248, 252. Freeman also explains here that Medows and he, riding on properly suppled horses, "kept within six inches of each other's knees without crowding; as the most delicate aids *carried* the horse a trifle either the one way or the other" (Freeman, 250), thus achieving the controlled and quiet knee-to-knee riding sometimes required by a cavalry in close formation.

34. Nosworthy, introduction to Warnery, *Remarks on Cavalry,* xxvii, xxxii.

35. Hinde, *The Discipline of the Light-Horse,* 10; on Pembroke, 22, and see 154; 44, 50; 148. Much thinner than Hinde's work in both size and substance, Captain L. Neville's *A Treatise on the Discipline of Light Cavalry* (1796), nearly half of it

devoted to maneuvers, doubtless added material of value to the contemporary conversation on cavalry, but of less value to the history of written works about cavalry.

36. Saxe, *Reveries, or Memoirs upon the Art of War,* vii; Warnery, *Remarks on Cavalry,* 11, 41.

37. Bismarck, *Lectures on the Tactics of Cavalry,* 10, 5–6. When Nolan refers to "the intelligence of the age," as we shall see, he invokes the Romantic zeitgeist reflected in both Bismarck's and his own work. Nolan, too, defines the brilliant cavalry leader as possessing both a clear coup d'oeil and the ability to act rapidly and decisively on what it reveals. An officer's actions, it follows, "partake more of the inspiration of genius than of the result of calculation and rule," though the latter also are necessary (Nolan, *CHT,* xxxv). While Seydlitz exemplifies that genius, Nolan concedes that the scale of the modern battlefield precludes an individual taking in "at one glance the whole state of affairs at any one time during a battle," and therefore makes it "more difficult for a cavalry leader to achieve the same results as the Prussians did under Seidlitz" (*CHT,* 22). In Nolan's romantic vision, however, neither inspiration nor genius will be denied: "Genius alone can make the poet and the general. . . . The real general [seizes] on the openings given by the enemy, as if by inspiration; there is no hesitation, ways and means are never wanting" (*CHT,* 122).

38. Coulston, introduction to Nolan, *CHT,* xiii.

39. The use of "dash," or, more commonly, "dashing" to describe cavalry, cavalrymen, or cavalry actions, ubiquitous in the nineteenth century and common in references to Nolan, survived intact into the twentieth century. See, for example, Denison, *HC,* 415; Goldman, *With General French and the Cavalry in South Africa,* 406, 412; Boniface, *The Cavalry Horse and His Pack,* 15; Childers, *WAB,* 11, 51, 295, and *GIBC,* 9, 63; and French, *Good-Bye to Boot and Saddle,* 49.

40. Anglesley, *A History of British Cavalry,* 2:83, 92; see Moyse-Bartlett, *Nolan of Balaclava,* 223–52. A letter from an unidentified officer at Balaklava, cited by Dawson, indicates that the Nolan hagiography extended to his horse: "I saw, quite distinctly, the horse of the late Captain Nolan standing over his body, licking his face, for a full hour. All who saw the noble beast felt it quite sensibly" (Dawson, *Real War Horses,* 126). Accounts of the battle and charge are legion; see, for example, Anglesey, 2:59–109.

41. Paget, quoted in Moyse-Bartlett, *Nolan of Balaclava,* 229; Chichester, "Lewis Edward Nolan," 41, 97.

42. Nolan, *TCRH,* 1. In his subsequent *Cavalry,* Nolan adds, "I do not assert that M. Baucher's system is faultless. I practised it for years, applied it to many hundred horses, and was myself obliged to make some trifling alterations to adapt it to the use of cavalry soldiers. . . . But what I assert is, that the system is the right *one:* it is founded on *reason* and *common sense,* not on immemorial custom and prejudice. So convinced was I of this by experience, that I wrote out and published the lessons as I had carried them out with hundreds of remount horses, to assist those who might be at a loss how to proceed when young horses join the regiments" (Nolan, *CHT,* 99).

43. Nolan, *TCRH*, 1–2, 5, 16, 50. Thus, for Nolan, "the 'Pirouette' [is] the most useful 'Air of the Manége' for a cavalry soldier; for, when engaged sword in hand with an enemy, he can turn his horse right, and left, and about, in an instant, and thus gain the advantage over his antagonist" (Nolan, *TCRH*, 19; and see 30–31). Nolan notes, "Monsieur Baucher includes the 'Piaffer' in his lessons for cavalry," but adds with regard to his own discussion of the air: "I think the quieter a cavalry horse is kept the better, and have, therefore, not included it in my drill, but merely add it, for the instruction of those who wish to carry out the system further than I consider necessary for cavalry purposes" (Nolan, *TCRH*, 42n). With respect to airs above the ground, Nolan regards the ballottade and capriole as "both equally useless" (Nolan, *CHT*, 92n). Lieut.-Colonel G. W. Key notes in his endorsement of Nolan, "It is not to be expected that every dragoon is to be a perfect riding master, or that his horse should be of the *haute école;* but it is greatly to be desired that every dragoon should be able, to break his own horse, to have him under thorough control, and to ride him with confidence and pleasure" (Key, letter in Nolan, *TCHR*, ix).

44. Coulston, introduction to Nolan, *CHT*, xii, xxxiv; Nolan, *CHT*, xxxvi, 60, 64 and 107, 117, 37.

45. Nolan, *CHT*, 90, 96, 97. "The difference between a school rider and a real horseman," Nolan writes, "is this: the first depends upon guiding and managing his horse for maintaining his seat; the second, or real horseman, depends upon his seat for controlling and guiding his horse" (Nolan, *CHT*, 95). On combining *manège* and cross-country equitation, DiMarco observes, "The most successful cavalry through the Napoleonic period were Frederick's cuirassiers under Seydlitz, who combined the discipline and technique of classical high school manège riding with a daring and aggressiveness that embraced cross-country riding at speed" (DiMarco, *War Horse*, 267).

46. Arguably the best known of several interwar American films celebrating British imperial might, *The Charge of the Light Brigade* followed *The Lives of a Bengal Lancer* (1935) and preceded *Gunga Din* (1939), the latter also based on a jingoist poem. The climactic battle scene of *Light Brigade* provoked substantial controversy related to animal welfare rather than to historical accuracy. The director Michael Curtiz used tripwires that either killed outright, or caused to be put down from wounds, upwards of a hundred horses. As a direct result, Congress passed legislation on animal safety in cinema and the ASPCA filed suit resulting in the banning of tripwires. Though Curtiz and Flynn, the latter an accomplished horseman, would make more movies together, the incident also resulted in Flynn's lifelong enmity toward the director.

47. Baker cites Nolan at 2n, 39n, 68, 87, and quotes a lengthy passage from Nolan at 24n–27n. Baker also shared with Nolan the infamy of public scandal, though one far seamier and more criminal. Baker's affair, indeed, resulted in his trial, incarceration, resignation of commission, and removal from the army, followed, as in a Victorian novel, by redemption and reinstatement prior to death. Brought to trial in 1875 following a serious incident with a Miss Dickinson, Baker

"was acquitted of attempted rape, convicted on the lesser charge of indecent assault, fined 500 pounds, and sentenced to a year's imprisonment" (see Anderson, "Baker, Valentine," and Jastrzembski, "The Reinvention of Valentine Baker").

48. Baker, *The British Cavalry*, 1. "Long-term calls for reform of the British cavalry during and after the Crimean War," Dawson writes, "were limited to a vocal minority—and the death of Nolan at Balaklava left Captain Valentine Baker (10th Hussars) and Major Francis Dwyer as the lone voices in the wilderness" (Dawson, *Real War Horses*, 130, and see 27). Scholars diverge, incidentally, on whether the Crimean War or the American Civil War was "the first modern war." Dawson notes, for example, "The Crimean War was the first modern, industrial war not just from its scale . . . but from the technologies involved: the electric telegraph, rifled muskets and artillery, land mines, ironclads, military railways in the combat zone and coverage of the events by the mass media" (Dawson, *Real War Horses*, 100). DiMarco, by way of contrast, notes, "The American Civil War was the first war where the *levee en masse* and nationalism of the Napoleonic period combined with the science of the industrial revolution" (DeMarco, *War Horse*, 231).

49. Baker, *The British Cavalry*, 10, 16. While breeding thoroughbreds for racing does not pose an inherent problem, Baker suggests, "Racing . . . has now fallen into the hands of a class of men who care nothing for the horse, provided that he but answers their purposes of speculation. . . . The weakly conditioned horses thus produced, eventually become the sires of our hunters and cavalry troop-horses" (Baker, 7, 9). A devotee of the horses of South Africa, Baker proposed that crossing thoroughbreds with "Cape horses of great bone and substance, would produce a small compact animal, unrivalled for light cavalry" (21).

50. Baker, *The British Cavalry*, 50–51, 53. "The saddle is one of the most important items in a dragoon's equipment," Baker notes (Baker, 48), but the current military saddle, in addition to encouraging an outdated seat, is "a complication of movable straps and buckles, pilches, woofs, &c., all made to take to pieces, and put together, like a Chinese puzzle" (Baker, 57). Nolan advocated, "That saddle is best for cavalry which, being of a simple construction, brings the soldier close to his horse, in a firm and easy seat" (Nolan, *CHT*, 85).

51. Denison, *MC*, v, xviii, 10–11, 59–61. In *A History of Cavalry*, Denison would repeat verbatim this encomium to the "murderous effect" of the revolver (Denison, *HC*, 422).

52. Denison, *HC*, 415, 416, 418, 420, 433.

53. Denison, *HC*, ix, ix–x, x, 426, x.

54. As the title page indicates, *Notes on Cavalry Service* was published in the series Military Manuals, described in an advertisement following the text: "A Series of Practical Hand-books on Military Subjects, written by Eminent Officers in the various Branches of the Army, and Illustrated with numerous Diagrams." Russell opens his preface, "The following Notes are based upon the latest regulations published for the English army, the works of trustworthy military authors, and the known opinions of experienced officers" (Russell, *Notes on Cavalry Service*, iii).

55. Russell, *Notes on Cavalry Service*, 7, 9, 39, 41, 46. Rimington would echo this last point four decades later: "The cavalryman must learn that never is the difference between cavalry and infantry to be observed more than when cavalry are acting dismounted" (Rimington, *Our Cavalry*, 187).

56. Chenevix-Trench, *Cavalry in Modern War*, 37, 53. Like Russell's *Notes on Cavalry Service*, Chenevix-Trench's *Cavalry in Modern War* also was published in a series, namely, as the title page indicates, Military Handbooks for Officers & Non-commissioned Officers.

57. Chenevix-Trench, *Cavalry in Modern War*, 121, 168. Badsey concurs with Chenevix-Trench's assessment of cavalry firepower: "An early nineteenth-century cavalry carbine was an altogether inferior weapon to an infantry rifle (whether muzzleloading or breechloading), but from the 1860s onwards carbines came into service with which dismounted cavalry stood a reasonable chance of out-shooting infantry at closer ranges, and by the 1880s there was little to distinguish a carbine such as the British Martini-Henry from the equivalent infantry rifle except the length of the barrel" (Badsey, *Doctrine and Reform in the British Cavalry*, 15).

58. Chenevix-Trench, *Cavalry in Modern War*, 179, 188. DiMarco makes an important point, however, when he writes, "Almost none of the lessons of the American Civil War cavalry experience had an impact on European cavalry. . . . The nineteenth century . . . saw great change but also conservative inflexibility in cavalry forces" (DiMarco, *War Horse*, 255, 268). Writers such as Chenevix-Trench, Haig, Childers, Ingelfingen, and Bernhardi may have studied those lessons, but the theories derived from them informed actual practice in limited ways. As the American officer Wagner wrote, to take one example, "American raids surpassed all previous operations of the kind, and have as yet been unequaled. . . . Ignoring the cavalry traditions of the Old World [Stuart] originated a new method of using mounted troops, and may be said to be the father of the cavalry tactics of the present day" (Wagner, *OT*, 212, 216). While the British officer Rimington later acknowledged, "The very idea of a cavalry raid is attractive and carries with it a certain romance," he also added, "The other side of the question may be seen in some of the unsuccessful raids entered upon by both sides in the American War, when raids became 'the fashion'" (Rimington, *Our Cavalry*, 145, 148). Traditionalism and conservatism aside, the United States and Europe differed in matters of politics and economics, demography and topography, and industry and technology, or, in short, in the *contexts* that informed cavalry theory and practice.

59. Schmidt died soon after his service in the war and on the commission that produced the Cavalry Regulations of 1873. The posthumous *Instructions*, compiled by Schmidt's long-time adjutant Captain von Vollard-Brockelberg, as their translator Captain C. W. Bowdler Bell explained in an introductory note, comprised "MS. notes dating from 1850, and the numerous Orders and Circulars which [Schmidt] had issued at different periods from the time when he commanded a squadron until within a few months of his death in 1875" (Bowdler Bell, introductory note to Schmidt, *Instructions for the Training, Employment, and*

Leading of Cavalry, iii). Ingelfingen followed his active service in the war with three works: *Ueber Kavallerie, Ueber Artillerie,* and *Ueber Infanterie* (1885–86), published in English as *Letters on Cavalry, Letters on Artillery,* and *Letters on Infantry* (1887–89). As Frederic Maude would note in 1903, Ingelfingen's later *Conversations on Cavalry* (1897) "contains the whole history of the movement of which the present German cavalry is the outcome. This book should be read in conjunction with General von Schmidt's work, which has been translated and supplied to all our regiments . . . for many of the difficulties with which von Schmidt had to grapple are not intelligible without a knowledge of the actual state of things with which he had to deal" (Maude, *CPF,* 230).

60. Schmidt, *Instructions for the Training, Employment, and Leading of Cavalry,* 1; 2, and see 187; 14; 121; 186. The conservatism of Schmidt's position is part and parcel with its pragmatism. As he concludes, "It cannot too often be repeated that the main thing is to carry out the mission in hand *at any price;* if possible this should be done mounted and with the *arme blanche,* but should that not be feasible, then we must dismount and force a road with the carbine" (Schmidt, 188).

61. Ingelfingen, *Letters on Cavalry,* 244; 6; 123, repeated on 137; 124; 127; 236; 244. Ingelfingen calculates that advances in the range of artillery and rifles means, "as a rule, cavalry must remain at a distance of two miles from the enemy if they are to be held halted out of action. . . . They have then to ride two miles before they can actually charge the enemy. This distance will probably be increased by the roundabout way by which the cavalry will have to advance. . . . The total distance which they will have to traverse . . . will thus amount to about four and a half miles. Even if at a gallop . . . it will be eighteen minutes before the cavalry reach the enemy" (Ingelfingen, 71–72).

62. Livingston and Roberts, *War Horse,* 27; Ottevaere, *American Military Horsemanship,* 8–9. Regulations for this new arm, of course, followed: *A Complete System of Tactics for Cavalry* (1834) and *A System of Tactics, adapted to the Organization of Dragoon Regiments* (1841), "almost entirely a direct translation and copy of the French *Regulation of 1829*" (Ottevaere, 11). Very early in the Civil War, Secretary of War Simon Cameron released the publication of the comprehensive *Cavalry Tactics or Regulations for the Instruction, Formations, and Movements of the Cavalry of the Army and Volunteers of the United States* (1862), by Brigadier General Philip St. George Cooke. Originally published in two volumes in 1860, Cooke revised his work and consolidated it into one volume for the Government Printing Office edition of 1862. This work "having been approved by the President," Cameron cautions, "all additions to or departures from the exercise and manoeuvres laid down in the system are positively forbidden" (Cameron, preliminary note to Cooke, *Cavalry Tactics,* v).

63. See Garrard, preface to *Nolan's System for Training Cavalry Horses.* Nolan's "System for Training Cavalry Horses," Garrard explains, "based upon the principles of equitation discovered by Monsieur Baucher, of France . . . is now out of

print and to preserve to the Cavalry Service so valuable a 'System,' this book has been prepared for publication. It is essentially the same as the original with the addition of a chapter on 'Rarey's Method of Taming Horses,' and one on the subject of 'Horse-Shoeing,'" both of these also reprints of earlier publications (Garrard, 1). *Nolan's System* carried a publication date of 1862, but Garrard's preface is dated October 1, 1861, or six months after the Battle of Fort Sumter. McClellan's *European Cavalry* carried a publication date of 1861, as opposed to 1862, but its publisher's preface is dated October 8, 1861, or a week after Garrard's preface to Nolan's *System*.

64. Publisher's preface to McClellan, *European Cavalry*, 3–4; McClellan, 15. *European Cavalry* carries the subtitle *Including Details of the Organization of the Cavalry Service Among the Principal Nations of Europe*. McClellan devotes nearly a hundred pages to the Russian cavalry, some seventy pages to the Prussian cavalry, under fifty pages each to the Austrian and French cavalries, and only a handful of pages to the British cavalry, noting, "The English tactics being easily obtained, and copies of it being in the possession of the War Department, it is deemed unnecessary to give extracts from it" (McClellan, 206).

65. Nolan and Garrard notwithstanding, McClellan reports that in both the Prussian and French military schools for equitation, "the Baucher system . . . has been tried and found to be unfit for cavalry purposes" (McClellan, *European Cavalry*, 135). He later adds, "I took especial pains to make inquiries, in relation to the Baucher system, of the cavalry officers of all the countries which I visited, and . . . the reply was uniformly the same: that is, that certain parts of the system . . . were good, and could be applied with advantage by individual officers to their own horses, but that the system would never answer for general introduction in the service" (McClellan, 199).

66. Brereton, *Educating the U. S. Army*, xii–xiii. Wagner explains the obvious though often unstated reason for the necessity of academic study in the preface to the first edition of *Organization and Tactics:* "The best school for acquiring a knowledge of organization and tactics is that furnished by actual experience in war. . . . Fortunately for the happiness of the human race, such schools of perpetual warfare do not exist" (Wagner, *OT*, viii). Consequently, officers and planners, especially those not yet tested in war, must study histories of previous wars but can test their theories only in future wars. Wagner's two textbooks, also adopted for use at West Point, enjoyed long-lasting influence: *The Service of Security and Information* was published in nine editions between 1893 and 1903, and the first edition of *Organization and Tactics* was followed shortly by four or more additional editions. As Brereton notes, "These books served as the basis for army postgraduate education until 1910" (Brereton, *Educating the U. S. Army*, xiv).

67. Wagner, *SSI*, 15–16; *OT*, 55, 60, 57. Wagner later elevates his point about cavalry independence to one of his five summary conclusions: "Cavalry unable to deliver effective dismounted fire action is essentially a dependent arm" (Wagner, *OT*, 237–38).

68. "When we compare the enormous results wrought by the American cavalry in the War of Succession with the feeble service rendered by the German
cavalry in France a few years later," Wagner cautions, "it is impossible to avoid the
conclusion that there was something radically wrong in the tactics and arms of
European cavalry" (Wagner, *OT,* 233).

69. Ottevaere, *American Military Horsemanship,* 25.

70. Carter, *HSB,* iv, 5–6. As acting chief of staff to the secretary of war, Carter
later wrote the memo preceding the text of *Notes on Equitation and Horse Training*
(1910), translated from a French manual used at Saumur and "published for the
information of the Regular Army and the Organized Militia" (War Department,
Notes on Equitation, 3).

71. Carter, *HSB,* 9, 48, 51.

72. Boniface, *The Cavalry Horse and His Pack,* xvii, 22, 53, 93, 97, 99. "To a
large extent," Boniface opens his preface, "any book on Cavalry is a compilation;
it is very difficult, if not impossible, to write anything absolutely original on the
subject, and, consequently, originality is not claimed for this work" (Boniface,
xvii). Totaling 538 pages and comprising 28 chapters, Boniface's compilation also
includes some 200 illustrative photographs, drawings, and diagrams. It carried the
full title, *The Cavalry Horse and His Pack, Embracing the Practical Details of Cavalry
Service, For the Use of Officers and Non-commissioned Officers of Cavalry.*

73. Carter, *HSB,* 1; Boniface, *The Cavalry Horse and His Pack,* 25; see McClellan, *European Cavalry,* 54. As John Sweetman writes in *Cavalry of the Clouds,* "In
October 1917 the British Prime Minister, David Lloyd George, proclaimed air-
men 'the cavalry of the clouds. High above the squalor and the mud . . . they fight
out the eternal issues of right and wrong. . . . They are the knighthood of this
war. . . . They recall the old legends of chivalry'" (Lloyd George, quoted in Sweetman, *Cavalry of the Clouds,* 9).

74. Wheeler-Nicholson, *Modern Cavalry,* 7.

75. Badsey, *Doctrine and Reform in the British Cavalry,* 4.

76. Wood, *Achievements of Cavalry,* vi. Wood served with distinction in the
Crimean War, Zulu War of 1879, and First Boer War of 1880–81; he was awarded
the Victoria Cross in the Indian Mutiny of 1857 and was appointed commandant
of Aldershot in 1889. "Those who wish to recall what cavalry has done in the
past," Field Marshal Sir John French would write, "should read and reread 'The
Achievements of Cavalry,' by Field-Marshal Sir Evelyn Wood, one of the very few
soldiers in the Army who has taken part as a combatant in European warfare. Sir
Evelyn Wood's war record probably surpasses that of any other officer in the
Army" (French, preface to Bernhardi, *CWP,* 16).

77. Wood, *Achievements of Cavalry,* 241, 247, 249–50. Wood's racist distinction between "Savage" and "European" warfare—the former evidently referring to
Afghanistan, India, and the colonies in general—finds its direct counterpart in the
equally racist American distinction between "Indian" and "American" protocols
for fighting in the Plains Indian Wars of the 1870s and 1880s, hypocrisy intact.

Though George A. Custer, for example, effusively praised the "wonderful skill in feats of horsemanship [of] the Indian warrior on his native plains," he described cavalry and Indians in one encounter as, respectively, "the representatives of civilized and barbarous warfare," and he later speaks of Indians as "the most cruel, heartless, and barbarous of human enemies" (Custer, *My Life on the Plains,* 199, 34, 274). As DiMarco points out, despite the American military's perception of Indian savagery, "The Civil War taught the army that the most decisive form of warfare was total war," practiced ruthlessly by Sherman both in the South and, subsequently, on the Plains—hence the army's brutal "large-scale winter campaigns" when Indians had to sacrifice mobility and occupy vulnerable permanent encampments (DiMarco, *War Horse,* 285).

78. Maude wrote some dozen books on military history and the introduction to a translation of Clausewitz's *On War* (1908, repr. 1918). Maude was neither dispassionate nor disinterested in his views. *Cavalry Versus Infantry,* for example, was published in the United States in an International Series of military works edited by Arthur L. Wagner, who cautioned in his editor's preface, "The views expressed by Captain Maude are original and fearless; and whether accepted in their entirety by the reader or not, they can not be perused without benefit by any thinking soldier" (Wagner, editor's preface to Maude, *CVI,* 8).

79. Maude, *CVI,* 30; 115; 185; 141; 145. As if to confirm Maude's idée fixe, the Prussian general Pelet-Narbonne would write in *Cavalry on Service* (1906), "We must learn to regard the life of a cavalryman as not so especially precious; we must expect the cavalry to suffer losses, when the situation requires it, just as much as the infantry do on many occasions, without being affected, and without making much ado about it" (Pelet-Narbonne, xi).

80. Maude opens his preface, "Since the following chapters were sent to the printers a new Army Order has been issued, bearing the date the 1st of March, 1903, directing that in future the British cavalry is to be taught to rely on 'the rifle, not the sword.' . . . This is by no means the first attempt which history records to develope fire-power at the expense of shock," misguided attempts that resulted in "disastrous consequences" for hybrid forces (Maude, *CPF,* vii). Spencer Jones clarifies: "In October 1900, lances and swords were officially withdrawn from regular cavalry regiments. An order from Lord Kitchener's headquarters added, 'The rifle will henceforth be considered the cavalry soldier's principal weapon'" (Jones, *From Boer War to Great War,* 174).

81. Maude, *CPF,* vii–viii, 2, 267, 273–74. Writers distinguished cavalry from both mounted riflemen and mounted infantry primarily on grounds, one, that the horse served cavalrymen as transportation and weapon, while serving mounted infantry only as transportation, and two, that cavalrymen were trained both for charging with *arme blanche* and for fighting dismounted with fire power, while mounted infantry were trained only for dismounted action. Wagner distinguishes "true dragoons" from mounted infantry on the same criteria (Wagner, *OT,* 58). Badsey notes, "While there was no comparison between the often expert

horsemen of some mounted rifle units and a mounted infantry unit improvised simply by giving an infantryman a horse, the terms 'mounted infantry' and 'mounted rifles' were often used loosely and interchangeably when discussing cavalry, resulting in sometimes unnecessary confusion" (Badsey, *Doctrine and Reform in the British* Cavalry, 14). Goldman, evidently an exception, did make such a distinction (Goldman, *With General French and the Cavalry in South Africa*, 415).

82. Earlier journalistic accounts of cavalry in the Crimean War set the generic conventions, as it were, for those on the Boer War written a half century later: British cavalry defends the long and righteous reach of the empire against the forces of hostile powers or feckless colonials; those adversaries prove more formidable than expected, however, and the cavalry suffers setbacks; but, in the end, British adaptability, mettle, and "cavalry spirit" prevail. The stories from the two wars also shared a dark theme: "a systematic breakdown in equine management" resulted in massive and unnecessary equine death, public outcry, and, in the latter case at least, reform in horse management (see Dawson, *Real War Horses*, 128).

83. Goldman, *With General French and the Cavalry in South Africa*, xii, vii–x, 405, 406, 409–10. Though an advocate for fire action, Goldman opposed discarding "the sword or lance" because he rejected the theory that "all chance of shock action for cavalry in the future has for ever disappeared" (Goldman, 413). J. G. Maydon made essentially the same case, even more emphatically, in *French's Cavalry Campaign* (1901): "Many writers challenge the Cavalry on the ground that the Mounted Infantry will supersede it. My eyes have seen that what the latter can do, the former can do better. The cavalry rides better, nurses its horses more, understands their needs better, and has a generally higher value than the infantryman on horseback. And, making allowance for the difference of arm, shoots not less well" (Maydon, *French's Cavalry Campaign*, ix).

84. The title page of *Cavalry Studies* identified then Major-General Haig as late inspector-general of cavalry in India. The *London Gazette*, November 22, 1907, reported his promotion from director of military training to director of staff duties at headquarters. A field marshal in the Great War, Haig succeeded Sir John French as commander-in-chief of the British Expeditionary Force.

85. Haig, *Cavalry Studies*, v–vi. For a contemporary discussion of staff rides, see Colonel R. C. B. Haking, *Staff Rides and Regimental Tours* (1908). Haking classifies the staff ride, as a training method, under "practical instruction on the ground without troops in peace" (Haking, *Staff Rides*, 1). As he elaborates, "It has been found by experience at the Staff College and elsewhere that officers learn more rapidly, and remember what they learn more easily, by means of a Staff Ride, or a Tactical Exercise on the ground, than by any other form of instruction. These exercises were first made use of in the British Army as a means for instructing officers at the Staff College in the correct way of applying the principles of strategy and tactics to a definite situation presented by a scheme, and also for teaching them the proper method of reconnoitring ground for strategical, tactical, and administrative purposes" (Haking, *Staff Rides and Regimental Tours*, 7–8).

86. Haig, *Cavalry Studies,* 19. Haig argues that "the decisive and governing factor" in cavalry success or failure is leadership, and that "a Cavalry Leader of any rank . . . needs mental power and capacity not often called on in Leaders of similar rank in the other arms" (Haig, *Cavalry Studies,* 7). Skeptics who believe that "the day of Cavalry is past" because of improved infantry arms are wrong: "Cavalry is in jeopardy [only] when it has no Leaders who understand how to train it" (Haig, 8, 15). Staff rides and reports, in short, both hone the skills of leaders and inculcate the values of leadership.

87. Haig, *Cavalry Studies,* 6, 18, 4. Haig writes that an officer leading a reconnaissance patrol has to "form a conclusion as to what the enemy is doing" and report it to his commander. This, however, "requires not merely the power of close observation. . . . The officer must have an understanding of the phenomena which come before his eyes" (Haig, *Cavalry Studies,* 1–2). Thus, he concludes his introductory chapter, "In the following pages an attempt is made to put before the reader [that is, officer] various situations with the object of accustoming him quickly to make up his mind, and then to interpret his decision into clear orders which can be easily executed" (Haig, 19). Ingelfingen, Wrangel, Loir, and Rimington each make the same point about the need to understand as well as to observe (Ingelfingen, *Letters on* Cavalry, 126–27; Wrangel, *The Cavalry in the Russo-Japanese War,* 65; Loir, *Cavalry,* 10; Rimington, *Our Cavalry,* 132–33).

88. Crowe, *Problems in Manoeuvre Tactics,* v; Gough, introduction to Monsenergue, *Cavalry Tactical Schemes,* xiii–xiv.

89. I presume Captain Loir to be (later) General Maurice-Eugène Loir (1870–1936), though he is identified only as Captain Loir, XX Army Corps Staff, on the book's title page and in its preface. An example of a type of work using actual events, reported factually, for purposes of instruction in applied knowledge, *Cavalry on Service* also illustrates the type of work, often in translation, published by Hugh Rees in the prestigious Pall Mall Military Series—a series that included, for example, Haig's *Cavalry Studies* and Monsenergue's *Cavalry Tactical Schemes.*

90. Loir, *Cavalry,* 123, 37, 91. "But there are extremists," Loir writes, "who consider it dishonourable, if a case should occur, to sheathe the sword in order to use the carbine, and there are also many others . . . who admit as a dogma that the essential arms for cavalry are the rifle, the machine gun, and field guns, and would hand up sabres and lances amongst trophies of ancient weapons" (Loir, 123).

91. Pelet-Narbonne, *Cavalry on Service,* vii–viii; Wrangel, *Cavalry in the Russo-Japanese War,* 55.

92. "In this book," Rimington writes in a one-sentence preface, "no attempt has been made to produce an exhaustive treatise on Cavalry; it has been written principally for junior officers of all arms" (Rimington, *Our Cavalry,* n.p.). A similar if more basic work, Major W. J. R. Wingfield's *Lectures to Cavalry Subalterns of the New Armies* (1915) comprised "lectures [for] subalterns of a Reserve Regiment of Cavalry . . . designed to teach them the rudiments of their work, in so far as it could be taught in theory as distinct from practice" (Wingfield, 5). In *An Officer's*

Notes (1917), finally, Captain R. M. Parker, U. S. Cavalry, also produced a "small pocket manual collecting in one volume the essential parts of military information required by a young officer [of all arms] before he can properly perform his duties" (Parker, 1).

93. Rimington, *Our Cavalry,* 78, 2, 158, 203, 208. Jones writes that Rimington was "generally considered the best horse-master in the British Army" (Jones, *From Boer War to Great War,* 198).

94. Jones, *From Boer War to Great War,* 170, 176. Bernhardi's *Cavalry in Future Wars,* translated by Charles Sydney Goldman, carried an introduction by French, and his *Cavalry in War and Peace* carried a preface by French; Childers's *War and the Arme Blanche,* carried an introduction by Roberts, and his *German Influence on British Cavalry* only a preface by its author. Childers opens the preface to *German Influence,* "This essay is meant to be read in connection with the facts and arguments adduced in [*Arme Blanche*]" (Childers, *GIBC,* iii). As Childers recalls the sequence in *German Influence:* "My book, 'War and the *Arme Blanche,*' was published in March, 1910, a month before the publication in England of his [Bernardi's] own second work, 'Cavalry in War and Peace,' whose consideration we have just concluded. In the course of the summer of 1910 the General published a series of articles in the *Militär Wochenblatt* criticizing my book, and those articles were translated and printed in the *Cavalry Journal* of October, 1910" (Childers, *GIBC,* 186).

95. Bernhardi, *CFW,* 3; *CFW,* 11, and see *CWP,* 39; *CFW,* 62, 184, 197. Rimington makes the same point about practical horsemanship: "Under the old Canterbury system much time was spent with a view to showing up a good ride of *haute école* animals, whilst the new system aims at training a horse which will go well in the ranks, and will be generally useful on a campaign, either in single combat or for a scout's riding, or for work in the ranks" (Rimington, *Our Cavalry,* 196).

96. Childers, *WAB,* 1. For Bernhardi, French, and "our own Cavalry school," Childers writes, the steel weapon "remains the weapon *par excellence* for the Cavalry, the indispensably decisive factor in inter-Cavalry combats, which are to take the form of shock duels," and this, he cites Colonel Henderson, despite "the revolution wrought in all modern tactics by the deadly efficacy of the smokeless, long-range magazine rifle" (Childers, *WAB,* 12–13, 15). As Jones summarizes, "The hybrid cavalryman that emerged in the years preceding the First World War was a conciliatory yet practical solution to the New School versus Old School debate that had followed the Boer War. . . . However, the hybrid concept did not draw universal admiration and was savaged in the notorious 1910 work *War and the Arme Blanche* by Erskine Childers" (Jones, *From Boer War to Great War,* 185). Jones adds, however, that Childers and others "who felt that cavalry should become purely mounted riflemen . . . were on the fringes of the debate" (Jones, 193). Even Lord Roberts, in his introduction to *Arme Blanche,* distances himself from those fringes: "I am driven to the conclusion that . . . all attacks can now be carried out far more effectually with the rifle than with the sword. . . . At the same

time I do not go so far as the author [Childers] in thinking that the sword should be done away with altogether" (Roberts, introduction to Childers, *WAB*, xii).

97. Childers, *WAB*, 8–9; *GIBC*, 27; *WAB*, 21–22, 36, 37, 319 and 323. In addition to his stated argument, Childers weaves a dissonant motif through *Arme Blanche* whose import becomes clear only in *German Influence*. Invoking the American Civil War and Boer War as proofs for the obsolescence of *arme blanche* and ascendance of firepower, he refers to the former conflict as "the really great and stimulating Anglo-Saxon precedent" (*WAB*, 52); he comments on the latter conflict, "To the Colonials . . . who were deeply imbued with the Boer belief in the rifle, the *arme blanche* was probably little more than a [British] race tradition" (*WAB*, see 55); and he exhorts his fellow Britons, "Study your own great war . . . this is *your* experience" (*WAB*, 354, and see 371). Childers also dismisses Bernhardi, "who writes not for Englishmen, but, as a German reformer. . . . We have nothing to learn even from him in the matter of Cavalry combat" (*GIBC*, iv). This becomes his stick for beating "British Cavalrymen, headed by Sir John French [who] acclaim the works of [Bernhardi] as the last word of wisdom on the tactics and training of modern Cavalry" (*GIBC*, 4). Childers's true bête noire, "far more reactionary" than even Bernhardi (*GIBC*, 5, 73, 77), "French refuses to read through British eyes the plain moral of the [Boer War] for Cavalry . . . throws the dearly-bought experience of his own countrymen to the winds . . . runs to foreigners who have no relevant experience for corroboration of an outworn creed, [and] gratuitously courts the same humiliation" (*GIBC*, 191).

98. Gray notes in his preface, "I intend to follow this work, Part I, with a work on 'Troop Leading of Cavalry,' wherein the decisions are based on the principles herein illustrated" (Gray, *Cavalry Tactics*, 4). Part 1 is indicated on the title page; part 2 never appeared.

99. Gray, *Cavalry Tactics*, 3; see 25, 49; 25. Gray's many references to "cutting down" and "running through" adversaries recall a running debate of scores of years, namely, the relative merits and demerits of cutting and thrusting. Gray reports, "Conversations with veterans of the Civil war lead to the conclusion that the enlisted man is going to use his saber almost entirely as a cutting weapon. . . . If, then, our best trained cavalry is going to use the saber chiefly as a cutting weapon, it is better to give it [cavalry] a weapon adapted to the manner in which it will be used rather than one adapted to the use of a trained fencer. . . . It is not to be inferred that the cut is to be preferred to the thrust, since the contrary is the case, but it is reiterated that the cut has been and will be used to the almost entire exclusion of the thrust" (Gray, 26). See note 11 above.

100. Hayne, *Lectures on Cavalry*, 28, 48, 54. Hayne writes, "The Germans do not favor such 'raids' [against lines of communication]," adding, "Bernhardi disagrees with the German Regulations" on this point, as on others (Hayne, 46–47). In *Germany's Fighting Machine* (1914), Ernest F. Henderson took a different view: "It is considered not unlikely that such 'raids' will play a great part in the present war. The Germans use the American word for the maneuver" (Henderson, 44).

101. *Notes on Infantry, Cavalry and Field Artillery,* "Reprinted for Officers Training Camps," comprises three lectures: "Notes on Infantry," by Major H. B. Fiske, "Notes on Cavalry," by Eltinge, and "Notes on Field Artillery," by Captain A. B. Warfield, "delivered to class of provisional second lieutenants, Fort Leavenworth" in January–March 1917. Assuming that the three lecturers, representing the three arms, were addressing the same audience, the series reflects a modern doctrine of increased cooperation among three more-or-less independent arms by promoting, among other means, the benefits of officers in one arm becoming familiar with the organization and operations of the other two.

102. Eltinge, "Notes on Cavalry," 32, 33, 34, 48. "The British used their mounted units as a highly mobile reserve which would ride to a threatened portion of the front and then dismount and fight as infantry," DiMarco writes, making clear that the practice still made use of cavalry's fundamental asset, mobility (DiMarco, *War Horse,* 310). Eltinge, in effect, responds *no* and *no* to the "frequently asked" questions "whether cavalry can attack mounted at all in face of the fire of modern firearms, and whether mounted infantry that used the horse for transportation but fought on foot would not do as well." He argues, one, that mounted charges by *large* bodies of horse now may be impossible, but not those by smaller bodies; and two, "In the work of the cavalry screen, in the pursuit and on sudden unexpected contact with any force, mounted work is quicker, safer and more effective than dismounted work" (Eltinge, "Notes on Cavalry," 45).

103. Spaulding, *Notes on Field Artillery,* 7.

104. As Major General Leonard Wood wrote in an introduction, Huidekoper (1874–1940) intended to inform "all Americans . . . who desire to replace our past haphazard policy by one which will be adequate to secure a reasonable degree of preparedness without in any way building up a condition of militarism" (Wood, introduction to Huidekoper, *The Military Unpreparedness of the United States,* xiii). Like the British, Huidekoper argued not for a standing army at wartime establishment, but for a military prepared to reach that scale quickly on an outbreak of hostilities. He warns, for example, that even if the current peacetime establishment was expanded substantially, U. S. Cavalry "would still be many thousands [of men and horses] short of what it ought to number" (Huidekoper, 472). Carter (1851–1925) concurred with the problem of defense that Huidekoper identified, but proposed a more radical solution. "Our isolation and the rivalries of European nations," he argues, "have served in the past to guard us from the usual results of neglect of an established military policy" (Carter, *AA,* 6), but conditions have changed, and the country now is vulnerable to invasion. While the United States should not become "a nation-in-arms, *it is an imperative duty that our military resources shall be organized and nationalized*" (Carter, *AA,* 12). To preserve peace, in short, the United States must put "our peace establishment upon a proper basis for expansion automatically in war," a goal achievable only through centralized policies (Carter, *AA,* 17). Woodhull (1843–1921) proposed a more specific military strategy and program. "The ocean no longer presents a barrier to invasion,"

Woodhull concurs with Carter, and cavalry would play an important role in meeting one (Woodhull, *West Point in Our Next War,* 45). Therefore, one, the army should equip cavalry with machine guns and train cavalry officers in their effective use (Woodhull, 74–77); two, "The cavalry . . . should be concentrated in one command . . . and should manoeuvre and fight independently of, *but in co-ordination with* the infantry" (189); and three, the army need look no further than the War between the States for tactical lessons (190).

105. Atteridge, editor's note to Bernhardi, *CFW,* 5. In postwar American popular culture, the German film director and actor, Erich von Stroheim, was known as the "man you love to hate" in reference not to the actor himself, who lived and worked in the United States, but to his frequent screen roles as an aristocratic German militarist.

106. Bernhardi, *GNW,* 9. Comparing and contrasting German and British bellicosity and propaganda, in addition to courting false equivalence, extends beyond our purview. It is worth noting, though, that Edmund von Mach, in his manifesto *What Germany Wants* (1914), contends not only that "Germany is not the home of militarism" (Mach, 122), but also that Bernhardi has no exclusive claim to militarist bombast—a point that he seeks to demonstrate with an appendix, "Quotations from the British 'Bernhardi,'" comprising comparably bellicose citations from Homer Lea's *The Day of the Saxon* (1912). For a German view opposite to Mach's and Bernhardi's, see Karl Liebknecht's widely read socialist polemic, *Militarism* (1917).

107. Bernhardi, *GNW,* 14, see 128, 198–99. Bernhardi had written in the 1910 edition of *Cavalry in War and Peace,* "So in the most modern war the cavalry remains the principal means of reconnaissance. Its activity may indeed be supplemented by airships, but will never be replaced by them." Atteridge, the editor of the 1914 abridged edition, added a lengthy footnote on advances in military use of aircraft for reconnaissance in the intervening years, though adding, as Bernhardi also believed, that aircraft would complement mounted cavalry rather than replace it (Atteridge, in *CWP,* 27, 27–28n).

108. Bernhardi, *HGMW,* xiv; see 29, 99; 71; see 96, 97; 103; 22; 86. Bernhardi's comment in his introduction, "England is particularly hostile towards us, in addition to France" (*HGMW,* xxii), follows directly from his epilogue to *Germany and the Next War,* where he had conjured "a real offensive and defensive alliance, aimed at us, between France and England," and urged using the "respite we still enjoy for the most energetic warlike preparation" (*GNW,* 283, 287). The British editor of *How Germany Makes War,* affirming the importance of Bernhardi's book "as an exposition of the ideas underlying the German plans for the war with the Allies," also cautions, "[Bernhardi] is writing as a leader of German military opinion for German readers, and looks at matters from a standpoint hostile to ourselves. As we read his words we must remember this" (editor's preface to *HGMW,* v–vi).

109. Badsey, *Doctrine and Reform in the British Cavalry,* 25. See Wrangel (*The Cavalry in the Russo-Japanese War,* 77–79), Monsenergue (*Cavalry Tactical Schemes,*

235), Hayne (*Lectures on Cavalry,* 17–18), Woodhull (*West Point in Our Next War,* 74–77), and Bernhardi (*CFW,* 178); Bernhardi stepped up his advocacy for machine guns in his postwar treatise *The War of the Future* (1921) (Bernhardi, *WF,* 135), and Wheeler-Nicholson advocated for "the formation of [cavalry] machine gun troops and squadrons" in *Modern Cavalry* (1922) (Wheeler-Nicholson, 110–12).

110. Kenyon traces the origin of the myth that machine guns mooted cavalry action to the official *History of the Great War* (1915–1945), compiled under direction of Sir James Edmonds, as did Terraine and Badsey before him, as Kenyon acknowledges (Kenyon, *Horsemen in No Man's Land,* 3). Still promulgating the myth in 1982, Alexis Wrangel described the machine gun as "the deadly weapon that was, more than all other ordnance, to spell the doom of cavalry" and therefore as "that great enemy of cavalry" (Wrangel, *The End of* Chivalry, 105, 107). More recently, though, Badsey has rejected the myth as an "enduring impression of cavalry charges" (Badsey, *Doctrine and Reform in the British Cavalry,* 31); Dawson has called it "the misleading image of the folly of the Great War" (Dawson, *Real War Horses,* 185); and both Kenyon and, after him, Macdonald have debunked it (Kenyon, 3; Macdonald, *Horses in the British Army,* 161). Kenyon points out, moreover, that the failure of some scholars "to examine fully the relationship between mounted soldiers and machine guns also leads to a failure to appreciate the further point that this new technology was applied as often *by* the cavalry as *at* them. . . . The machine gun was at least as much the friend of the mounted soldier as his enemy" (Kenyon, 6). As he later concludes, "This firepower [Vickers and Hotchkiss machine guns], combined with their mobility, made the cavalry a potent force on the battlefield" (Kenyon, 231).

111. Bernhardi, *WF,* 125, 127, 133–34. In the same year, *Tactics and Technique of Cavalry* (1921), published by the U. S. General Services Schools, observed, "The doctrine of making a cavalry that would be effective mounted or dismounted . . . to its maximum requires that the cavalry be armed not only with weapons for mounted combat, but also with weapons for dismounted attack which will put it on a par with foot troops. It is for this reason that machine rifles and machine guns have been added to the armament of our cavalry" (General Service Schools, 14–15).

112. Another controversial cavalryman, Wheeler-Nicholson published *Modern Cavalry* in the same year that he faced court-martial for insubordination and, soon after, resigned from the military. As reported in the *New York Times* of February 6, 1922, Wheeler-Nicholson, "who risked court-martial for subordination yesterday in going over the heads of his superiors to send a letter direct to President Harding alleging that 'Prussianism,' favoritism and inefficiency ruled in the treatment of junior officers in the army, said tonight he had made the charges, regardless of consequences to himself, for the 'good of the service.'" A reviewer in the *New York Times* of December 15, 1940, savaged Wheeler-Nicholson's later book, the polemical *Battle Shield of the Republic* (1940), as "critical, in fact, almost abusive," and "guilty of many vitriolic generalities and minor inaccuracies," though the reviewer also granted that it "nevertheless should help to stimulate much-needed reforms in the Army."

113. Wheeler-Nicholson advocated the idea, shared by other writers, that "the solution" to instilling the importance and virtue of horsemanship in a trooper "is to assign a man a horse, the animal to be his as long as he cares for it properly," or, as he later rephrases it, "to cultivate a high degree of personal responsibility and liking for his mount in each individual soldier of cavalry" (Wheeler-Nicholson, *Modern Cavalry,* 68, 74). Wheeler-Nicholson's point about dressage finds a parallel in his point about equitation: "[The trooper] should be taught more individual care of his horse and a little less of the refinements of riding than he now receives" (Wheeler-Nicholson, 69).

114. Liddell Hart, "ACW," 414, 417, 417–18.

115. Liddell Hart, *RMA,* v; "ACW," 418; *RMA,* 10, 16, 17.

116. Liddell Hart, *RMA,* 63, 69. As enthusiastic about the tank as Liddell Hart, H. G. Wells had written in *War and the Future* (1917), "The Tank is only a beginning in a new phase of warfare. . . . The decisive factor in the sort of war we are now waging is the production and right use of mechanical material; victory in this war depends now upon three things: the aeroplane, the gun, and the Tank developments" (Wells, 168). Less sanguine with the benefit of hindsight, Kenyon indicts the polemicists in "the post-war debate over mechanization," including Liddell Hart, who promulgated a mythic "victory of the tanks" and "the [horse] cavalry as a metaphor for the obsolescent, in contrast to new 'modern' methods of warfare" (Kenyon, *Horsemen in No Man's Land,* 3, 8). Debunking that myth, Kenyon argues, "Far from being a replacement for the cavalry, tanks had at best a complementary rôle on the battlefield. . . . Thus the relationship between the two is far more complex than simply the 'new' taking over from the 'old'" (Kenyon, 15). The job of the tank, Kenyon cites Brigadier General Hobart, "was to crush the wire, to allow the infantry into the German defences, and to destroy machine gun positions. . . . Thus in no sense did the appearance of tanks render the cavalry obsolete" (Hobart, quoted in Kenyon, 242–43). In addition to promoting the tank, Liddell Hart touted gas, implausibly, as a "humanizing" weapon: "Gas may well prove the salvation of civilisation from the otherwise inevitable collapse in case of another world-war" (Liddell Hart, in the chapter "The Humanity of Gas," *RMA,* 110).

117. Brereton, *The Horse in War,* 34; Coulston, introduction to Nolan, *CHT,* xxiv. Early accounts of cavalry exploits on the Eastern Front include Lieut.-Colonel R. M. P. Preston's *Desert Mounted Corps* (1921) and *Lt.-Col. R. Evans's Brief Outline of the Campaign in Mesopotamia* (1930). In an introduction to the former, Harry Chauvel, Lieut.-General, late Commanding the Desert Mounted Corps, writes, "If [Preston's book] does nothing else, it must demonstrate to the world that the horse-soldier is just as valuable in modern warfare as he ever has been in the past" (Chauvel, introduction to Preston, *The Desert Mounted Corps,* vii). Among recent scholars, Jones, for example, writes, "Cavalry did not become extinct in the First World War; indeed, they achieved a number of notable successes, particularly in the Middle East" (Jones, *From Boer War to Great War,* 205); Brereton writes that France provided cavalry few opportunities to prove its worth in mounted action, but "Palestine showed

that, given the conditions, the horse—and the *arme blanche*—could still play a decisive role in modern warfare" (Brereton, *The Horse in War,* 133); and DiMarco writes, "From [October 1914] on, the impact of mounted forces on the western front was negligible," but cavalry "played an important role on the eastern front" (DiMarco, *War Horse,* 310). Badsey, however, mounts a convincing counter argument for the case, "But just as it is an exaggeration to call the Palestine campaign 1917–1918 a 'cavalry war,' so it is a historical error to omit the cavalry's contribution to the Western Front" (Badsey, *Doctrine and Reform in the British Cavalry,* 16).

118. Brereton, *The Horse in War,* 147; DiMarco, *War Horse,* 334, 309. Basic works on horse cavalry in the period, particularly German cavalry, include Piekalkiewicz, *The Cavalry of World War II* (1976), Richter, *Cavalry of the Wehrmacht, 1941–1945* (1997), and Dorondo, *Riders of the Apocalypse: German Cavalry and Modern Warfare, 1870–1915* (2012).

119. See, for example, Livingston and Roberts, *War Horse,* 36–37. As DiMarco notes, moreover, "War horse and rider stepped into the twenty-first century in the fall of 2001, when the United States deployed Special Operations Forces to Afghanistan" (DiMarco, *War Horse,* 352). A friend from the intelligence community, reading this chapter in draft form, noted that commanders now sit in office chairs rather than saddles, finger keyboards, toggles, and mice rather than reins, and dispatch reconnoitering patrols of drones rather than horses and men.

120. Truscott, *The Twilight of the U. S. Cavalry,* xx; Coffman, foreword to Truscott, x.

121. Truscott, *The Twilight of the U. S. Cavalry,* 10, 84–85, 86, 86–87, 155. The first three officers were Colonel Bruce Palmer, whose goal was to develop "tactics and techniques, organizations, and methods of employment that would make maximum use of armored vehicles and motor transportation" (Truscott, *The Twilight of the U. S. Cavalry,* 95); Colonel Charles Scott, who, despite "his vast knowledge of horses and horsemanship . . . was also one of the most enthusiastic advocates of mechanization for cavalry" (Truscott, 97); and Major-General Herbert Crosby, for whom mechanization was not, contrary to popular opinion, "an enemy of Cavalry," but rather its "greatest friend"—one that "will create a greater demand for Cavalry than ever before" (102). The fourth was General John K. Herr, who "believed in *horse* cavalry " and fought "every form of mechanization at the expense of the horse cavalry at the time when all trends in military thought in all modern armies was toward mechanization and armor" (155).

122. French, *Good-Bye to Boot and Saddle,* 123. Badsey nicely unpacks the connection between "smartness" and social class: "The Army's unofficial hierarchy of smartness was headed by the Household Cavalry and the Foot Guards, after which the order of smartness between regiments was largely a matter of peer-group recognition. . . . Just as officers (except those commissioned from the ranks) were expected to be gentlemen by birth, status and lifestyle, so smartness involved the level of conspicuous display by a regiment's officers, and its ability to attract high-status officers and patrons" (Badsey, *Doctrine and Reform in the British Cavalry,* 7).

Badsey also adds, though, a critical point: "Wealth and gentlemanliness in officers, and smartness and patronage in regiments, were in this period fully compatible with military professionalism and competence as well as with bravery" (Badsey, 9).

123. Gough, introduction to French, *Good-Bye to Boot and Saddle*, 13. In French's view, to cite just one example, Cromwell may have been "boorish" and capable of "deeds of merciless inhumanity," as well as a regicide and a zealot, but he also displayed, "above all, a true understanding of the *cavalry spirit*. He rejoiced in leading cavalry charges, invariably riding home with the utmost gallantry and *élan* (French, *Good-Bye to Boot and Saddle*, 171–72). Sir John French and Field Marshal Edmund Allenby, "the last of our great cavalry leaders," exemplified that spirit (see French, 208–26).

124. French, *Good-Bye to Boot and Saddle*, 37; 227; 261, 267; 271. "My father," French writes, "who, at the time of the [*arme blanche*] controversy, was the foremost authority on cavalry in the country, maintained that, if deprived of their steel weapons, [cavalrymen] would lose the true *cavalry spirit*, and become reduced to the status of mounted infantry. . . . In winning his *casus belli* for retention of the *arme blanche*, and for the preservation of British cavalry in its traditional rôle, my father achieved a victory whose far-reaching effect was beyond computation, and rendered to the cavalry, the army, and the country an incomparable service" (French, *Good-Bye to Boot and Saddle*, 232, 234).

125. Wood, *Achievements of Cavalry*, 243; Denison, *HC*, 420; Ingelfingen, *Letters on Cavalry*, 8, 58.

4. Remounts and Wastage

1. Illustrating a common twentieth-century view, Brereton writes that "generals—who were mostly cavalrymen—continued to be obsessed with the dream of an ultimate 'cavalry breakthrough,' and vast numbers of troop horses were kept in the field awaiting the chance that never came" (Brereton, *The Horse in War*, 125). Rejecting a related myth "that the high command was dominated by cavalry officers and was thus, by extension, incompetent," revisionist historians such as Kenyon and Badsey demonstrate that high command, including Field Marshal Haig, quite reasonably looked throughout the war toward such a "breakthrough" (see Kenyon, *Horsemen in No Man's Land*, 3).

2. Quoted in Winton, *Theirs Not to Reason Why*, 353.

3. Boniface, *The Cavalry Horse and His Pack*, 39. See Galtrey, *The Horse and the War*, for the same use of the term "remount" in Great Britain.

4. Boniface, *The Cavalry Horse and His Pack*, 39; McClellan, *European Cavalry*, 111, 173, 126.

5. Boniface, *The Cavalry Horse and His Pack*, 1; Livingston and Roberts. *War Horse*, 28, 31.

6. Boniface, *The Cavalry Horse and His Pack*, 533. By comparison, as Boniface had noted, "The latest census of the horse supply of Great Britain shows a

total of about 3,000,000 . . . about 70,000 [of whom] are considered likely to be found fitted for military purposes" (Boniface, 47).

7. Boniface, *The Cavalry Horse and His Pack*, 40; see Bernhardi, *CFW*, 152–59. A few years before Boniface, General William H. Carter contended in *Horses, Saddles and Bridles* (1895), "Military horse breeding farms are enormously expensive . . . and are altogether inadequate to meet the demands of modern armies [and] it is safe to conclude that the horses required for public service [in the United States] will continue to be purchased from private breeding farms" (Carter, *HSB*, 4–5).

8. Rommel, *The Army Remount Problem*, 103, 108, 113; 114, and see Livingston and Roberts, *War Horse*, 47. As Rommel notes, "Congress has expressly forbidden the War Department to expend any of its appropriation for breeding purposes" (Rommel, 111), leading to the described plan devised jointly by the Agricultural and War Departments in 1910 and approved in 1911 (see Rommel, 119–21). Livingston and Roberts point out that the single sentence expressly forbidding use of appropriation for breeding purposes appeared in the Annual Appropriation Acts of 1909, 1910, and 1911, "but had not appeared" before 1909, "nor did it appear" after 1911 (Livingston and Roberts, 58). They also describe how the plan, described by Rommel at its inception and officially started in 1913, subsequently worked out in practice (60–62).

9. Borden, *What Horse for the Cavalry?* i, ii, v; 1, 8.

10. Reese, "Breeding Horses for the United States Army," 341, 350, 354, 355–56.

11. Dawson, *Real War Horses*, 4; Nolan, *CHT*, 216; Baker, *The British Cavalry*, 16. Baker elaborates his point and calls for government intervention in chapter 3, "Government Studs" (Baker, *The British Cavalry*, 16–22). An earlier point made by Baker explains part of his reasoning: "It is a pity that no Government remount establishment has been formed in the Persian Gulf. . . . The present mutiny has proved how lamentably deficient was our supply of remounts. The lack of cavalry has been the main cause of the protraction of the struggle" (11n).

12. An insular nation with a global empire, England paid great attention to the logistics of efficient and safe transoceanic shipping of horses. For two mid-nineteenth-century examples, see Arthur Shirley, *Remarks on the Transport of Cavalry and Artillery: With Hints for the Management of Horses, Before, During, and After a Long Sea Voyage* (1854), or Baker, chapters 23, "Embarkation of Horses," and 24, "Treatment on Board Ship," in *The British Cavalry* (1858). Bounded by two oceans, the United States also attended to the sea transportation of horses. See, for example, McClellan's discussions of the policies and practices in France (McClellan, 186–89) and England (208–11), in *European Cavalry* (1861); Carter, chapter 16, "Transportation of Horses by Rail and at Sea," in *HSB* (1895); Dyer, chapter 8, "Transportation," in *Handbook for Light Artillery* (1896); and Boniface, chapter 13, "Transporting Cavalry Horses," in *The Cavalry Horse and His Pack* (1903).

13. Goldman, *With General French and the Cavalry in South Africa*, 17, 433, 442. "A purchasing officer," Goldman writes, "should have experience of horses on

active service. The ordinary knowledge of hunters, polo-ponies, and racehorses may assist him; but it falls far short of what he really requires. . . . This was especially noticeable in the animals secured for service with the artillery, many of which were supplied by the omnibus companies" (Goldman, 434–35).

14. In *Small Horses in Warfare* (1900), Gilbey advocated the breeding of small, quick, durable horses, "that useful type which is suitable for light cavalry and mounted infantry," like those ridden by the Boers in South Africa (Gilbey, *Small Horses in Warfare*, 1). He later adds, "For artillery and transport, however, we shall always need powerful horses, and the draught power required is only to be obtained with height" (Gilbey, *Small Horses in Warfare*, 34). *Small Horses in Warfare* was a companion to *The Great Horse* (1889, reissued 1900), Gilbey's history of the Shire breed in war and peace.

15. Gilbey, *Horse-Breeding in England and India*, 4, 8; 11, 59. "Since the South African War," Gilbey points out, "the War Office authorities have been bombarded with schemes and suggestions . . . for increasing the home-bred supply of Remounts" (Gilbey, *Horse-Breeding in England and India*, 7). Discussing plans for "Horse Breeding in India," Gilbey refers to a *Report of the Horse and Mule Breeding Commission* (1901) that explored the viability of breeding cavalry horses and mules in India rather than importing them from Australia (Gilbey, 52–65; and see Government of India, *Report of the Horse and Mule Breeding Commission*).

16. Gilbey, *Horses for the Army*, preliminary note to 1913 edition, n.p.; 2, 11, 15. "The Remount Depôts," Gilbey explains, "would serve purposes distinct from the Registration and Subsidy Schemes; the former would give the sorely needed stimulus to British horse-breeding and provide the Army with its peace requirements of a much better class than it now enjoys; the Registration and Subsidy Schemes provide against the contingency of war" (Gilbey, *Horses for the Army*, 33).

17. Haig, *Cavalry Studies*, 4. H. G. Wells disparaged horse-mounted cavalry both during and after the war, on the grounds, as he wrote in *War and the Future* (1917), that "there has been a colossal buying of horses for the British army, a tremendous organisation for the purchase and supply of fodder, then employment of tens of thousands of men as grooms, minders and the like, who would otherwise have been in the munitions factories or the trenches" (Wells, 154). Kenyon has debunked this belief as "the 'fodder' myth" (Kenyon, *Horsemen in No Man's Land*, 3).

18. Preston, *The Desert Mounted Corps*, 325. As Anthony Dawson explains, "Mechanisation of transport, especially after 1914, meant that certain types of horse fell out of use and their trade collapsed. Principally affected were the light draught horses used to pull horse buses and trams but which were highly prized by the Army for its own transportation purposes" (Dawson, *Real War Horses*, 13). The same was true in the United States.

19. Wheeler-Nicholson, *Modern Cavalry*, 30.

20. Macdonald, *Horses in the British Army*, 41; Ottevaere, *American Military Horsemanship*, 67; Livingston and Roberts, *War Horse*, 36; Baker, "The Only

Mounted Unit in the U. S. Army," 13–14, and W. F. B. "The 287th M. P. Horse Platoon—In Memorium," 20; R. L. B., "The British Army's Last Mules," 59.

21. Carter, *HSB,* 209. In the American Civil War, for example, Carter writes that the Union army "required more than 500 horses each day for remounts; and this is the measure of destruction of horses during the same period [1863–64]" (Carter, *HSB,* 211). Regarding the British in Crimea in 1854–1855, he adds, "All sorts of excuses have been made for the losses in this campaign, but the melancholy fact remains that the horses were starved to death. During a period of six months the loss of transport horses was thirty-eight per cent., and out of 5048 cavalry and artillery horses there remained at the opening of spring 2258" (Carter, *HSB,* 218). Goldman cites as an example from the Boer War, "The waste of horses in the three cavalry brigades between [February 15 and February 27 in 1900] was 1,474" (Goldman, *With General French and the Cavalry in South Africa,* 441). And M. F. Rimington, the debacle of South Africa on his mind, cautioned, "The care of the horse is the weak link in the cavalry chain. . . . Strong measures are needed to counter-act our daily growing ignorance of horsemastership" (Rimington, *Our Cavalry,* 208). Among recent scholars, Dawson writes, "The death of so many horses [in the Crimean War] was due to a systematic breakdown in equine management" (Dawson, *Real War Horses,* 128); and DiMarco extends the critique to the American Civil War: "Effective horse management was the exception among Union forces rather than the rule. . . . In the early years of the war, ignorance at all levels of command led to the negligent if not cruel treatment of Union army horses. . . . Throughout the war hundreds of thousands of horses were lost due to neglect or mistreatment" (DiMarco, *War Horse,* 244–45). With respect to South Africa, DiMarco adds, "British horse management in the field was . . . as derelict as it had ever been" (DiMarco, 301). Moreover, "Intentionally poor ration allocations resulted in the near starvation of thousands of horses that later succumbed to disease. British commanders, despite veterinary advice to the contrary, mandated a near-starvation ration for unit horses. This was because of poor appreciation of the type of horses that the troops had, and the working conditions of the horse" (304).

22. See Goldman, *With General French and the Cavalry in South Africa,* 433–48; Rimington, *Our Cavalry,* 18, and see Wheeler-Nicholson, *Modern Cavalry,* 72.

23. Chenevix-Trench, *Cavalry in Modern War,* 95; Denison, *HC,* 437; Maude, *CPF,* 186; see Schuessler, "The Horse of World War I," 24, and Matha, *General Chamberlin,* xviii–xix.

24. Winton, *Theirs Not to Reason Why,* 94; 58, 118. Though generally consistent, estimates differ. Soon after the Boer War, for example, Erskine Childers reported, "The total number [of horses] provided for the British army was 518,794 (mules 150,781). The net wastage accounted for was in horses 347,007 (mules 53,339)" (Childers, *WAB,* 277n). More recently, DiMarco writes that "350,000 of 500,000 perished" (DiMarco, *War Horse,* 302), and Dawson that "of the 518,000 horses sent to South Africa, 326,000 died" (Dawson, *Real War Horses,* 184); Brereton refers to "350,000 dead out of a total of 520,000 remounts sup-

plied" (Brereton, *The Horse in War,* 118). Winton states the problem succinctly: "It is impossible to arrive at a single figure for the total number of animals supplied, or used during the war" (Winton, *Theirs Not to Reason Why,* 90)—a problem that also applies to numbers regarding wastage and disposal. A memorial to Horses in the Second Boer War was erected in Port Elizabeth, South Africa, in 1905 (see Wikipedia, Horse Memorial).

25. The imputation of poor horsemastership was not unique to the Boer War. Brereton writes of the Napoleonic Wars, "On service, the heaviest horse wastage was usually caused by privation and poor horsemastership rather than by enemy action" (Brereton, *The Horse in War,* 76). Advocating the establishment of remount depots, Gilbey noted that, as a collateral benefit, they would provide venues for teaching and learning "the elements of the horse master's business—a science in which it is to be feared we are sadly deficient" (Gilbey, *Horses for the Army,* 32).

26. Winton, *Theirs Not to Reason Why,* 101. Winton's assessment of horsemastership in the artillery has strong consensus: DiMarco cites Anglesey on this point (DiMarco, *War Horse,* 304), and Anglesey, in turn, had cited Major-General Frederick Smith (Anglesey, *History of British Cavalry,* 4:357n). Kenyon offers the positive note, "The other key area of improvement resulting from the South African experience was in horse-mastership. . . . By 1914 this situation had changed radically and the care and management of his mount was one of the key skills of any British cavalryman" (Kenyon, *Horsemen in No Man's Land,* 19).

27. Bernhardi, *HGMW,* 8; Kenyon, *Horsemen in No Man's Land,* 242; Winton, citing the Horse Census, *Theirs Not to Reason Why,* 252.

28. Winton, *Theirs Not to Reason Why,* 145; Dawson, *Real War Horses,* 188. Writing during the war under some sanctions, Galtrey wrote in *The Horse and the War* (1918), "I am permitted to say that actually 165,000 horses were impressed in the United Kingdom in the first twelve days of the war" (Galtrey, 16).

29. See Winton, *Theirs Not to Reason Why,* 354, 357, 371.

30. As Brereton notes regarding the Western Front, for example, "all but super-heavy 'position' artillery was horse-drawn; there were six horses to each field gun, eight or twelve to the medium and heavy types, and each gun had its supporting ammunition wagons. It was reckoned that on any sector of the front between 1916 and 1918 there was one gun for every ten yards of the line; the number of horses involved is incalculable" (Brereton, *The Horse in War,* 126).

31. Travis, *The Mule,* 37–38; Galtry, *The Horse and the War,* 51; Travis, 44.

32. Winton, *Theirs Not to Reason Why,* 371. These actions, of course, generated atrocity propaganda, such as that retailed by D. S. Tamblyn in *The Horse in War* (1932): "It is a well-known fact that the hunnish instinct of the German submarine commanders caused them to have pictures taken of these ghastly scenes, which were exhibited in Germany during the war. This will go down, very far down into history as one of their hellish atrocities" (Tamblyn, 19). For a contemporary overview of U. S. shipping of military horses at the beginning of the twentieth century, see Bonifice, chapter 13, "Transporting Cavalry Horses," *The Cavalry Horse and His*

Pack, 268–98; for an overview of British shipping in the Great War, see Galtry, chapter 6, "The Crossing Overseas," *The Horse and the War,* 54–60. For a recent overview of shipping and its dangers in the Great War, see Winton, "The Maritime Contribution and Submarine Menace," *Theirs Not to Reason Why,* 367–71; and for a broad overview of British military shipping 1750–1950, see Macdonald, chapter 5, "Transporting Horses by Sea," *Horses in the British Army,* 50–65.

33. Brereton, *The Horse in War,* 115. Brereton elaborates: "The twentieth century brought in motorised transport and the heavier-than-air flying machine. Methods of warfare were revolutionised by the infantry's automatic machine guns capable of firing 800 rounds a minute, with a range of 2,000 yards; their rifles were magazine-fed, accurate to more than a mile, and the new 'quick-firing' field guns enabled the artillery to put down devastating concentrations with 12 and 15 1b high-explosive shells" (Brereton, *The Horse in War,* 115).

34. Moore, *Animals of the Great War,* 151. In the memoir *Russian Hussar* (1965), Vladimir Littauer recalls this battlefield incident: "At that time a heavy shell burst about thirty feet from our group, which was now watching the charge. Surprisingly, none of us was hurt, although a dozen horses standing some 500 feet away were killed or wounded" (Littauer, *Russian Hussar,* 194).

35. Fairley, *Horses of the Great War,* 76; see Dyer, *Handbook for Light Artillery,* 49, and Galtrey, *The Horse and the War,* 50; Goldman, *With General French and the Cavalry in South Africa,* 459.

36. The estimate of eight million comes from the website of The Animals in War Memorial (see "Animals in War Memorial," The Royal Parks); the estimate of three million of six million from Winton (*Theirs Not to Reason Why,* 429). See Galtrey, *The Horse and the War,* 104.

37. Moore, *Animals of the Great War,* 151; Hemingway, *Death in the Afternoon,* 122.

38. Winton, *Theirs Not to Reason Why,* 419, and see 34. Macdonald notes, "At the end of the First World War, in addition to horses and mules, there were also 56,287 camels, bullocks and donkeys in army service" (Macdonald, *Horses in the British Army,* 173).

39. Brereton, *The Horse in War,* 138. Similarly, an order that surviving Australian "walers" (a sturdy crossbred horse) over the age of eight be destroyed and the younger sold "for local use" met strong resistance that caused the order to be rescinded: "The troopers preferred to give their mounts a quick and humane death rather than condemn them to a life likely full of pain and abuse which the troopers knew to be typical of working horses in Egypt" (DiMarco, *War Horse,* 323). French, *Good-Bye to Boot and Saddle,* 55.

40. For discussion of Dorothy Brooke and her rescue work, see Brereton, *The Horse in War,* 139–40; Fairley, *Horses of the Great War,* 133–34; and Winton, *Theirs Not to Reason Why,* 424–25.

41. For an introductory pictorial overview, see Butler, *The War Horses,* with 188 illustrations, varying widely in medium and genre, together with informative legends.

42. Matania's original illustration, in color, clearly displays the severity and bloodiness of the horse's wounds; the monochromatic version used on the American Red Star Animal Relief poster, while not hiding the wounds or blood, makes both less obvious and thus, while still pathetic, perhaps less unsettling to a potential donor.

43. Horses sometimes were outfitted with full-face gas masks and sometimes with the appliances resembling feedbags as shown in Figure 4. D. S. Tamblyn observed, however, "The use of horse gas masks was not practical, more especially under shell fire" (Tamblyn, *The Horse in War*, 38). Horses, moreover, were vulnerable to a greater danger than inhalation, since walking through muddy ground in the area of a gas attack, even days later, could produce large and painful blisters on their feet and legs. Under the heading, "Protection Against Gas," the U. S. military manual *Tactics and Technique of Cavalry* (1930) advises, "Animals . . . are more susceptible [than men] to the effects of blistering agents, such as mustard gas. . . . Animals suffer most from the effects of vesicants (blistering agents) which have been splashed upon them or collected on legs or belly in passing over or lying upon infected ground" (General Service Schools, *Tactics and Technique of Cavalry*, 351; and see Fairley, *Horses of the Great War*, 56).

44. Findley, *The Wars*, 57–72.

45. *A Concise Catalogue of Paintings, Drawings and Sculpture of the First World War 1914–1918* lists over five thousand pieces of Great War art, representing many subjects, in the Imperial War Museum alone. Countless other pieces are held in museums on the Continent, in Great Britain and North America, and elsewhere. Avant-garde artists included the German painters August Macke and Franz Marc, founding members of Der Blaue Reiter, who both died in the war.

46. Fairley, *Horses of the Great War*, xii.

47. In the memoir *My Horse Warrior* (1934), Lord Mottistone (General Jack Seely) reproduced a drawing of Munnings's portrait of Warrior with Mottistone aboard and recalled the occasion of its painting: "The background, at a distance of about three thousand five hundred yards, was German territory" (Mottistone, *My Horse Warrior*, 114–17).

48. In 1919, Kemp-Welch executed an important fourth contribution to Great War painting that was not equestrian in subject matter. The Empress Club of London commissioned Kemp-Welch to commemorate "Women's Work in the Great War 1914–1918." The finished painting, eighteen feet in height, was "installed at the Royal Exchange and unveiled by Princess Mary on 28 April 1924" (see Wortley, *Lucy Kemp Welch*, 130–32).

49. Cecil Aldin not only produced artwork for recruitment posters for the Women's Land Army (see Figure 2), but, as Fairley writes, "With the shortage of skilled horsemen [on the home front], Aldin was the pioneer in recruiting women into the Remount depots. Soon there were depots, like Russley Park, entirely run by women" (Fairley, *Horses of the Great War*, 122; and see Winton, *Theirs Not to Reason Why*, 312). Capable horsewomen, they trained remounts for a variety of

military uses, both draft and saddle. The United States Remount Service adopted this model of recruiting civilian horsewomen for training military horses at its depots. Commissioned by the Women's Work Section of the Imperial War Museum to commemorate the horsewomen of the Russley Park depot, Kemp-Welsh submitted *The Ladies' Army Remount Depot, Russley Park, Wiltshire, 1918*. The museum preferred *The Straw Ride,* with its imposing scale of roughly six by thirteen feet.

50. For the Animal War Memorial Dispensary, see "8 Memorials to Animals in the First World War," Heritage Calling. Dorothy Brooke's pioneering work continues to this day in the Brooke USA Foundation. See "100 Years On, USA's Million War Horses Honored," *Horse Talk.*

51. On the Animals in War Memorial, See "Animals in War Memorial," The Royal Parks; and "Animals in War Memorial—The Monument," Animals in War Memorial. On the War Horse Memorial, see "Ascot Unveils War Horse Memorial to Commemorate WW1," BBC News. On the memorial plaque at Saint Jude-on-the Hill, see "8 Memorials to Animals in the First World War," Heritage Calling. The United States paid its Great War horses similar homage. In "The Horse of World War I," Schuessler cites "the inscription appearing on the bronze memorial tablet commemorating the horses and mules which died during the World War erected in the State, War and Navy Building by the American Red Star Animal Relief, October 21, 1921: 'This tablet commemorates the service and sufferings of the 243,135 horses and mules employed by the American Expeditionary Forces overseas during the great war which terminated November 11, 1918, and which resulted in the death of 63,682 of those animals. What they suffered is beyond words to describe'" (Schuessler, "The Horse of World War I," 25).

52. Rankin, *A Subaltern's Letters to His Wife,* 43–44, 45, 50.

53. Galtrey, *The Horse and The War,* 109.

54. Galtrey, *The Horse and the War,* 13; Haig, prefatory note to Galtrey, 11; H. M. the King, cited in Galtrey, 13; Galtrey, 98, and see 19.

55. Galtrey, *The Horse and the War,* 13, 14, 20, 20, 24, 36, 43 and 122, 20, 34, 35, 43. Galtrey's call for the breeding of Percherons in England aligns with his praise of "American" horses: "If more were required to justify the Allies' splendid war-horse, it is the firm conviction . . . that the light draught of American origin has come to stay in this country. . . . For, apart from their value as war-horses, they must attract the employer of the general utility horse. After all, they are a distinct type. . . . By comparison the British light draught is a nondescript, a misfit. . . . But the Yankee was essentially and absolutely a light draught horse, true to type, varying not at all in character and very little in the non-essential details. . . . He simply must take his place, and an important one, too, in the horse population of these Islands" (Galtrey, 36). The title page of *The Horse and the War* identifies, as its American publisher, "The Percheron Society of America, Union Stock Yards, Chicago, U.S.A."

56. Tamblyn, *The Horse in War,* 9, 21, and see also, for example, 15, 26, 43.

57. The razor bomb "exploded about a foot above the ground," Tamblyn writes, "with the result that the bomb splinters would strike the legs and abdomen of our horses; in fact, in many instances our animals were completely disemboweled" (Tamblyn, *The Horse in War,* 32).

58. Tamblyn, *The Horse in War,* 35. Arguably, the opportunities for a horse's humane death in the Boer War were even fewer. Rankin wrote to his wife, "Even the merciful bullet that puts an end to pain was generally denied them. Leave has to be obtained to shoot an animal, and frequently, where large bodies of men are in the vicinity, such a course is attended with great danger" (Rankin, *A Subaltern's Letters to His Wife,* 49).

59. Tamblyn, *The Horse in War,* 56, 88.

60. Mottistone, *My Horse Warrior,* 47; 29, 51, 117; 142; see 72, 137, 151. Warrior neither refuses jumps nor bolts, *ever,* regardless of shell fire or other circumstances; he is affectionate and loyal, brave and bold, judicious and wise, prescient and preternaturally lucky; he is a natural leader who, in the charge, "set an example to all the other horses" (Mottistone, 89). Hence, perhaps, Mottistone's assurances, "Those who read this book may think that what I write is fantastic and untrue, but I beg them to believe it is not" (47; and see 125).

61. Mottistone, *My Horse Warrior,* 80; 108, and see 133; 134. For references to German courage, see 64, 94–95, 110, 134. The film, *Into the Storm* (2009), quotes Churchill's comment on sighting his friend Mottistone during the Boer War: "I saw him at the head of a column of British cavalry, riding twenty yards in front, on a black horse. I thought of him as the very symbol of British Imperial power" (O'Sullivan, *Into the Storm;* and see "End of Glory: 'Into the Storm,'" International Churchill Society). Whether the comment is apocryphal or not is unknown to me.

62. Johnston, *Riding into War,* 33. Johnston's memoir was published as volume 4 of The New Brunswick Military Heritage Series under the auspices of The New Brunswick Military Heritage Project.

63. Johnston, *Riding into War,* 55. On equine labor, Johnston recalls, for example, "On August 16, [1917], pack horses carried 160,000 rounds of small arms ammunition from the reserve dump to the battery positions. During the night of August 17–18, another 500,000 rounds were delivered" (Johnston, 45). On equine and human danger, "The ground all around here was one continual series of shell holes filled with water and to slip into one of them was quite serious. In the first place a lot of them were so big a man and horse would be drowned if he slid into one of them. The mud was so soft that it was almost impossible to get footing to climb out" (Johnston, 57). Death by drowning was not uncommon on the Western Front (see, for example, Hochschild, *To End All Wars,* 285–88).

64. Sassoon, *MFM,* 255.

65. Hemingway, *In Our Time,* 12. "Truly the horse might cry out more loudly than any other creature, 'Give peace in our time, O Lord,'" Mottistone wrote, most likely not alluding to Hemingway but rather invoking their shared source, the Book of Common Prayer (Mottistone, *My Horse Warrior,* 108).

66. Hemingway, bibliographical note, in *Death in the Afternoon,* 517.

67. Hemingway, *Death in the Afternoon,* 16, 20. Though bull and man can share the honor of bringing the tragedy to its climax, "The tragic climax of the horse's career has occurred off stage at an earlier time; when he was bought by the horse contractor for use in the bull ring" (Hemingway, *Death in the Afternoon,* 6). This is one aspect of Hemingway's baroque thesis that the death of the horse is not tragic, but rather "comic," or, more precisely, a "burlesque" of tragedy (6–7).

68. "At the first bullfight I ever went to I expected to be horrified and perhaps sickened by what I had been told would happen to the horses" (Hemingway, *Death in the Afternoon,* 1). Neither horrified nor sickened, he becomes his own case study of an aficionado who feels discomfort when seeing horses injured on the street or in the hunt, but not when seeing them injured in the bullring.

69. Hemingway, *Death in the Afternoon,* 4. Hemingway bases his distinction between the aficionado and the casual spectator on the principle of the organic unity of part and whole and of form and meaning in a work of art: the aesthetic and *moral* integrity of the work depend on it. On that basis, Hemingway postulates, "The tragedy of the bullfight is so well ordered and so strongly disciplined by ritual that a person feeling the whole tragedy cannot separate the minor comic-tragedy of the horse so as to feel it emotionally. . . . The aficionado, or lover of the bullfight . . . has this sense of the tragedy and ritual of the fight so that the minor aspects are not important except as they relate to the whole" (Hemingway, *Death in the Afternoon,* 9). A casual spectator who focuses on the horse's suffering, in contrast, exaggerates and isolates an integral part of the ritual from the whole. Failing to appreciate the ritual's organic unity, that spectator also fails to experience its effects.

70. Hemingway, *Death in the Afternoon,* 12, and see 184. As Hemingway writes in "An Explanatory Glossary," regarding use of the peto, "The frank admission of the necessity for killing horses to have a bullfight has been replaced by a hypocritical semblance of protection which causes the horses much more suffering but . . . saves the horse-contractor money, enabling the promoters to save money and allowing the authorities to feel that they have civilized the bullfight" (*Death in the Afternoon,* 467).

71. Hemingway, *Death in the Afternoon,* 2.

72. Hemingway, *A Farewell to Arms,* 196.

73. Hemingway, *Death in the Afternoon,* 501; *Manchester Courier,* quoted in Dawson, *Real War Horses,* 187.

5. Hunting in the Trenches

1. Childers, *WAB,* 290.

2. Fussell, *The Great War and Modern Memory,* 92. Sassoon's diaries were published posthumously as *Diaries, 1915–1918* (1983), *Diaries, 1920–1922* (1981), and *Diaries, 1923–1925* (1985). His trilogy of nonfictional memoirs

comprises *The Old Century and Seven More Years* (1938), *The Weald of Youth* (1942), and *Siegfried's Journey* (1946).

3. Sassoon, *MIO*, 458, 514; *SP*, 650.

4. Fussell, *The Great War and Modern Memory*, 82, 90, 92, 311.

5. Sassoon, *MIO*, 358, 362, 431, 435, 411, 420, 465.

6. Sassoon, *MFM*, 221; *MIO*, 464, 450, 457, 473; Graves, *Goodbye to All That*, 201; Sassoon, *MFM*, 221.

7. Sassoon, *SJ*, 203, see also *MIO*, 347, 382; *MIO*, 481, 387.

8. Sassoon, *MFM*, 225; Fussell, *The Great War and Modern Memory*, 74; Sassoon, *MIO*, 300; Sassoon, *SJ*, 97.

9. Sassoon, *MIO*, 396, 422, 423; Liddle Hart, *RMA*, 104. Sherston often employs metaphors of machinery and mechanization (see also *MFM*, 235, 247; *MIO*, 382; *SP*, 653). The *war* is a machine that kills men and processes their corpses. The *army* is a machine that not only kills soldiers efficiently but that also requires them to function as its moveable parts: "If we continue to accept war as a social institution we must also [teach] children in schools . . . to offer their finest instincts for exploitation by the unpitying machinery of scientific warfare" (*SP*, 655). Finally, the individual *soldier* is a machine made of interchangeable parts: "Wonderful chaps these medicos!" Graves writes. "They supply spare parts as though one were a motor-car" (*Goodbye to All That*, 208). Common in later anti-war literature and film, the trope of mechanization recurs, for example, in Oliver Stone's film *Platoon* (1986): "Now, I got no fight with any man that does what he's told," says the homicidal Sergeant Barnes, "but when he don't, the machine breaks down, and when the machine breaks down, we break down, and I ain't gonna allow that from any of you, not one." The twentieth-century trope of war as destructive machine extends and inverts the eighteenth- and nineteenth-century trope of army as productive machine (see chapter 3, note 31).

10. Sassoon, *MIO*, 327; *SP*, 640; *MIO*, 343, 409.

11. Sassoon, *MIO*, 452; *SP*, 525, and see 527; *SP*, 537; *SP*, see 537, 541, 654. For convalescing combatants, the war also was the nightmare from which there is no awakening, an "underworld of dreams [where] by night each man was back in his doomed sector of a horror-stricken Front Line, where the panic and stampede of some ghastly experience was re-enacted among the livid faces of the dead" (*SP*, 556–57).

12. Sassoon, *MIO*, 365, 392–93, 418; *MFM*, 257; *MIO*, 426–27.

13. Sassoon, *MIO*, 449; *SJ*, 119.

14. Wohl, *The Generation of 1914*, 119; Sassoon, *SP*, 655, and see 557. One of the most iconic and enduring legacies of Great War literature, the trope of the "lost" generation has a basis in reality. As Wohl testifies: "Junior officers suffered heavier losses than the men who served under them. . . . The younger the junior officer and the more privileged his education, the more likely he was to be killed. . . . Everything was done to keep the memory of *these* dead—that is, the dead of the elite—alive" (Wohl, *The Generation of 1914*, 115). Graves recalls: "At

least one in three of my generation at school died; because they all took commissions as soon as they could, most of them in the infantry and Royal Flying Corps. The average life expectancy of an infantry subaltern on the Western Front was . . . only about three months; by which time he had been either wounded or killed. . . . Flying casualties were even higher (Graves, *Goodbye to All That*, 52).

15. Sassoon, *MFM*, 230.

16. Sassoon, *SJ*, 76.

17. Fussell, *The Great War and Modern Memory*, 102; Sassoon, *MIO*, 421; *SP*, 548.

18. As a young man, Sherston recalls, "There was no doubt that I had a fondness for books—especially old ones. But my reading was desultory and unassimilative. I esteemed my books mostly for their outsides" (Sassoon, *MFM*, 79). Later, when Sherston was a new recruit, books served nostalgia rather than aesthetics: "Books about England were all that I wanted. I decided to do plenty of solid reading at the Army School" (*MIO*, 287)—service that intensifies at the front: "Late that night I was lying in the tent with *The Return of the Native* on my knee. . . . I wanted to explore the book slowly. It made me long for England, and it made the War seem waste of time" (*MIO*, 357). As a new antiwar activist on leave, "I had returned to Butley resolved to read for dear life—circumstances having made it imperative that I should accumulate as much solid information as I could . . . for the completion of my education as an intellectual pacifist" (*MIO*, 489). Despite his failure to become an "intellectual" (*SP*, 537), he continues to read: when in Palestine, Conrad, Tolstoy, and Hardy (*SP*, 590, 597, 601), and when back in France, Duhamel, de la Mare, and Whitman (*SP*, 617, 621, 628). Sadly, he comes to the realization: "One cannot be a useful officer and a reader of imaginative literature at the same time. . . . The mechanical stupidity of infantry soldiering is the antithesis of intelligent thinking" (*SP*, 627).

19. Sassoon, *MFM*, 104. The epigraph to *Sherston's Progress* reads: "I told him that I was a Pilgrim going to the Celestial City" (*SP*, 515), and Sherston speaks at least three times of his "pilgrimage" (*MFM*, 161; *SP*, 524, 563). He also leads a fatigue party that "would have made an impressive picture of 'Despair'" (*MIO*, 434), and has a nightmare that finds him "in some slough of despond" (*SP*, 555). In facing and surviving these trials, of course, Sherston marches toward salvation. Determining that his refusal to disavow his inflammatory statement "was an inevitable conjuncture in my progress, and that such temptations must be resisted inflexibly . . . I began to feel quite optimistic about the progress I was making" (*MIO*, 506–7). Sherston, finally, jokes that Cromlech's reverence for French brothels would lead you to believe "that for a good young man to go through Havre or Rouen was a sort of Puritan's Progress from this world to the next" (*MIO*, 386).

20. Sassoon, *SP*, 656.

21. Sassoon, *MIO*, 346, 443; *SP*, 640.

22. Holmes, *Tommy*, xvii; Wohl, *The Generation of 1914*, 93, 95. The dust wrapper of the first edition of Sassoon's collection *The Old Huntsman* (1917) describes the poems as "hall-marked with the impress of the new age opened to us by the war, its atrocities, and its heroisms."

23. Sassoon, *SJ*, 17, 40, 190. "The war poets," Wohl writes, "had provided the theme [for the memoirists and novelists]: doomed youth led blindly to the slaughter by cruel age. But a decade passed before this theme was developed in prose in a systematic or sustained fashion. Then came a rash of books about the generation of 1914 and their war experiences" (Wohl, *The Generation of 1914*, 105). Holmes adds, "And so *we* remember the war not as we might, through the eyes of 1918, as a remarkable victory so very dearly won, but through the eyes of 1928 as a sham which had wasted men's lives and squandered their courage" (Holmes, *Tommy*, xxiv). The phenomenon was not restricted to British memoirists or readers. As Holmes continues, *All Quiet on the Western Front* (1929), Erich Maria Remarque's now classic novel, "struck a powerful chord with many veterans looking back at the war from the deep disillusionment of the late 1920s, and in a sense more accurately reflects the state of its author and his friends in 1929 than the condition of the German Army twelve years before" (Holmes, *Tommy*, xix). Since three of the major poets (Blunden, Graves, and Sassoon) also wrote the first three major memoirs in the late 1920s (each its author's first book of prose), the theme identified by Wohl not surprisingly migrated from poetry to prose.

24. Wohl, *The Generation of 1914*, 106; Sassoon, *MIO*, 291, 425, 471.

25. Sassoon, *MFM*, 239; *MIO*, 501; *SJ*, 55.

26. Sassoon, *MFM*, see 252; *MIO*, 339; *SP*, 546, 543.

27. Sassoon, *SJ*, 55, 69.

28. Discussing the generic prototype, Goethe's *Wilhelm Meister's Apprenticeship*, Margaret Drabble describes its protagonist in terms that fit Sherston closely: "Wilhelm provides the model of the innocent, inexperienced, well-meaning, but often foolish and erring, young man who sets out in life with either no aim in mind or the wrong one. By a series of false starts and mistakes and with help from well-disposed friends he makes in the course of his experiences, he finally reaches maturity and finds his proper profession" (Drabble, *The Oxford Companion to English Literature*, 100).

29. Sassoon, *MFM*, 21, 176. Sassoon writes in *Siegfried's Journey*, "Short poems are my natural means of expression, and for sustained description I have come to prefer prose" (*SJ*, 71). With reference to *Fox-Hunting Man*, he adds that Edmund Gosse "urged me to undertake a long poem which would serve as a peg on which . . . my reputation could hang. He suggested that I might draw on my sporting experiences for typical country figures. . . . Ten years later Gosse's critical intuition proved to have been correct. The peg on which my popular reputation finally suspended itself was—though written in prose—essentially in accordance with his advice" (*SJ*, 100–01).

30. Sassoon, *MFM*, 15, 92, 146, 46, 47, 72; *MIO*, 317.

31. Fussell, *The Great War and Modern Memory*, 95.

32. Citing 177 examples, Anne Grimshaw estimates that books specifically on hunting published in England between the wars accounted for "25% of the total output of equestrian literature" (Grimshaw, *The Horse: A Bibliography of British Books*, 160).

33. Sassoon, *SJ*, 163; *MFM*, 229–30. Sherston ascribes to Hoadley Rectory, family home of a friend, "the charm of something untouched by modernity" (*MFM*, 122); he reports "little evidence of modernity in what we did and saw" when hunting in the midlands (*MFM*, 202); he also regards the growing population in the Packlestone country as "an intrusion" and would like to "clear every mean modern dwelling out of the hunt" (*MFM*, 203). With respect to country folk, Ladies Diana Shedden and Apsley advised readers of *"To Whom the Goddess . . ."*: "And indeed the hunting lady who takes no part in the life of the country-side and fails to know the real country people and their lives, their troubles, difficulties, and their joys and recreations, misses a great deal" (Shedden and Apsley, 271).

34. Sassoon, *MFM*, 69, 220.

35. Sassoon, *MFM*, 93, 95, 155; *SP*, 639.

36. Sassoon, *MFM*, 235; *MIO*, 447.

37. Sassoon, *MFM*, 133, 145.

38. The converse was also true. Recalling "an ideal platoon officer," Sherston adds, "I need barely say that he had never hunted. He could swim like a fish, but no social status was attached to that" (*MFM*, 248).

39. Sassoon, *MFM*, 246. Hunting could confer the same advantage, presumably, in the militia as well as in the regular army. As Graves notes, "The militia majors [were] for the most part country gentlemen with estates in Wales and no thoughts in peacetime beyond hunting, shooting, fishing, and the control of their tenantry" (Graves, *Goodbye to All That*, 179).

40. Sassoon, *MIO*, 383.

41. Sassoon, *MFM*, 88, 106, 232. British officers were expected to provide their own uniforms and kits, so Major W. J. R. Wingfield's training manual, *Lectures to Cavalry Subalterns of the New Armies* (1915), for example, carried advertisements for uniforms from Burberry's, waterproofs from Anderson, Anderson & Anderson, and sling map cases from Forster Groom & Co. Civilians and soldiers also could patronize army and navy stores, as do Aunt Evelyn (*MFM*, 142) and Sherston (*MIO*, 316).

42. Sassoon, *MFM*, 107; *MIO*, 406, 383, 407, 412.

43. Graves, *Goodbye to All That*, 56. The word "pluck" appears often as a virtue in sporting, equestrian, and cavalry writing of the period. In *Riding Recollections* (1878), for example, G. J. Whyte-Melville, cites "nerve" as "the one essential [for] proficiency in the saddle," adding that "nerve" and "pluck" are constituents of "valour," but that pluck is "a brilliant and imposing costume" while nerve is "an honest wear-and-tear fabric, equally fit for all weathers" (Whyte-Melville, *Riding*

Recollections, 113). Wildly popular, *Riding Recollections* had eight editions in its first two years of publication and many more in the following decades. Horses, too, needed pluck: "Pluck is the best quality in a horse," James Fillis wrote in *Breaking and Riding,* and "no amount of training can . . . give pluck to an animal which is always ready to shy or spin round" (Fillis, 35, 173).

44. Wohl, *The Generation of 1914,* 97.

45. Sassoon, *MFM,* 12, 13, 15, 86, 144, 147, 187. Sherston identifies fox-hunting as his "career" right after recalling that his attorney had chided him for failing "to take up some serious calling and occupation" (*MFM,* 86). With an inherited "net income of about six hundred [pounds] a year," and an annual allowance from it of 450 pounds (*MFM,* 72, 86), Sherston enjoys a steady and ample income without an "occupation."

46. Sassoon, *MFM,* 114, 135, 194. Sherston fears being inadequate not only as a hunt rider: "There were moments when I felt acutely conscious of the absolute nullity of my past as a race-rider" (*MFM,* 152).

47. Sassoon, *MFM,* 12, 196. Graves writes, "Once, I remember, an old major laid it down axiomatically that every so-called sportsman had at some time or other committed a sin against sportsmanship." For example, "a New Army major, a gentleman-farmer [confessed] that his estate had been over-run by foxes one year and, the headquarters of the nearest hunt being thirty miles away, he had permitted his bailiff to protect the hen roosts with a gun" (Graves, *Goodbye to All That,* 180).

48. Sassoon, *SJ,* 52, 57; *MFM,* 230; *MIO,* 294; *SP,* 645.

49. Sassoon, *MFM,* 99, 100, 100, 34; *MIO,* 476, 502. Cavalry treatises and manuals often carried cautions like this one from Nolan's *Cavalry:* "When a charge has once begun, carry it out whatever may be the odds that suddenly present themselves against you" (Nolan, *CHT,* 179), the implication being that indecision or wavering commitment during a charge invited danger or disaster. As General Haig observed in *Cavalry Studies,* "With the other arms it is possible to break off an action; not so with the Cavalry charge—fate must run its course" (Haig, 13).

50. One could portray, for example, master, huntsman, whippers-in, and hounds as analogous, respectively, to commander in chief, commanding general, field commanders, and troops (see Cayce, "Fox Hunting, Past and Present").

51. See Sassoon, *MFM,* 18, 25, 84, 138, 197, respectively, on Sherston's horses.

52. Sassoon, *MFM,* 223–24; *MIO,* 291. Once in France, Sherston takes "over the job of [company] Transport Officer. This was an anti-climax, for it meant that I shouldn't go into the trenches [but was] considered appropriate, on account of my reputation as a fox-hunting man" (*MFM,* 259).

53. Sassoon, *MIO,* 362, 336; Sassoon, *The War Poems,* 25; Graves, *Goodbye to All That,* 185.

54. Principal characters include Aunt Evelyn and Tom Dixon, Denis Milden and Stephen Colwood, Dick Tiltwood and David Cromlech, and W. H. R. Rivers, Sherston's therapist at Slateford War Hospital. Stephen and Tiltwood are killed in the war; Dixon dies of pneumonia in the war; "and Denis had become so

remote that I seldom remembered him" (Sassoon, *MFM,* 265). Too numerous to mention, "types" include the admirable Mr. Colwood, who "loved riding, shooting and fishing." He also married into money, "enabling his three sons to be brought up as keen fox-hunters, game-shooters, and salmon-fishers" (*MFM,* 123); and the formidable Mrs. Oakfield, "the feminine gender of a jolly good fellow . . . a fine figure of a woman . . . as she sailed over the fences in her tall hat and perfectly fitting black habit . . . a brilliant horsewoman [who] rode over the country in an apparently effortless manner" (*MFM,* 206). Though Sherston briefly invokes several women, he treats at length only Aunt Evelyn.

55. Sassoon, *MFM,* 108. Dixon, "like every good groom, would sit up all night with a hunter rather than risk leaving a thorn in one of its legs after a day's hunting" (*MFM,* 14), just as Mr. Buckman of the Coshford Vale Stag Hunt "when times were bad . . . would go without his dinner himself rather than stint his hounds of their oatmeal" (*MFM,* 146). Bill Jaggett, by contrast, "was a hulking, coarse-featured, would-be thruster; newly rich, ill-conditioned, and foul-mouthed" (*MFM,* 108), whose "boon companion Roger Pomfret was a good-for-nothing nephew of Lord Dunborough who blundered about the country on a piebald cob and vied with Jaggett in coarseness of language and general uncouthness" (*MFM,* 111). Note the class distinctions between the household staffer Dixon, landholding stag-hunter Buckman, nouveau riche Jaggett, and minor aristocrat Pomfret.

56. Sassoon, *MIO,* 426–27; *MFM,* 211; *MIO,* 427.

57. See Sassoon, *MIO,* 318 and 342, respectively, for soldiers bathing and a doomed young warrior; *MFM,* 41; on Milden, see *MFM,* 171–72; *MIO,* 445. Tiltwood enters the narrative (*MFM,* 241) almost immediately after "Denis had disappeared into a cavalry regiment and was still in England" (*MFM,* 228). Comically phallic as Dick Tiltwood's full name may be, even by Sassoon's standards, "the name *Dick,*" as Fussell observes of Sassoon and Graves, "was becoming conventional for this sort of thing" (Fussell, *The Great War and Modern Mem*ory, 208).

58. Sassoon, *MFM,* 17, 43, 45, 205–6, 211.

59. Sassoon, *MFM,*178; *MIO,* 430.

60. Sassoon, *MFM,* 128–29, 204; *MIO,* 451; *MFM,* 33; *MIO,* 344.

61. Sassoon, *MIO,* 378–79; 296, 509.

62. Sassoon, *MFM,* 218, 225; on amputation and atrophy, see *MFM,* 235 and 265, respectively; *MIO,* 318, and see *MFM,* 227, *MIO,* 457, *SP,* 654; *MFM,* 276; *SP,* 654. The antiwar activist Thornton Tyrrell (counterpart of the real-life Bertrand Russell) cautions Sherston that his act "will probably land you in prison," but also admonishes him not to be discouraged: "You will be more alive in prison than you would be in the trenches. . . . By thinking independently and acting fearlessly on your moral convictions you are serving the world better than you would do by marching with the unthinking majority who are suffering and dying at the front because they believe what they have been told to believe" (*MIO,* 479). Sassoon encapsulates in that one sentence the point of "Sherston's progress" and *Sherston's Progress.*

63. Sassoon, *MFM*, 174; *MIO*, 351, 360, 429, 326, 461, 419. "Dreamers" (1917) pictures soldiers in reverie: "I see them in foul dug-outs, gnawed by rats, / And in the ruined trenches, lashed with rain, / Dreaming of things they did with balls and bats"; and "Break of Day" (also 1917) captures a soldier, deep in a reverie of hunting while shivering in his dug-out, waiting for dawn and the signal to go over the top: "And hunting surging through him like a flood / In joyous welcome from the untroubled past; / While the War drifts away, forgotten at last." In "Together" (1918), though, the speaker, with hunting "all day" in mind, meta-phorically conjures the war dead: "I shall not think of him: / But . . . I know that he'll be with me on my way / Home through the darkness to the evening fire" (Sassoon, *The War Poems*, 88, 103, 115).

64. Sassoon, *MFM*, 225, *SP*, 607; *MIO*, 361, and see *MIO*, 398, *SP*, 567; *MIO*, 352. "I talked to [Dick Tiltwood] about fox-hunting," Sherston notes of their early friendship, "which never failed to interest him. He had hunted very little, but he regarded it as immensely important" (*MFM*, 241). Sherston also recalls "one morning when [David Cromlech] was shaving with one hand and reading *Robinson Crusoe* in the other. Crusoe was a real man, he remarked; fox-hunting was the sport of snobs and half-wits." This provokes an argument, of course, about the virtues and vices of hunting (*MIO*, 385).

65. Sassoon, *MIO*, 487; *MFM*, 255–56; Graves, *Goodbye to All That*, 242; Sassoon, *MIO*, 377, 380, 383.

66. Sassoon, *SP*, 562, 565, 584, 633–36.

67. Sassoon, *MFM*, 279; *MIO*, 303–4, and see Hochschild, *To End All Wars*, 285–87; Sassoon, *MIO*, 444. Sherston frequently refers to the practice of tun-neling in trench warfare, either to move men to the enemy's position, or to plant explosives to blow up the enemy's position. His lengthiest and most harrowing passage concerns the Hindenburg Outpost trench and tunnel (*MIO*, 433–46) and a simultaneous "underground attack along the Tunnel, as well as along the main trench up above" (*MIO*, 436). Associated with World War II, the term "foxhole" had limited use in the Great War. In *Undertones of War,* Blunden evokes "dead men in field-gray overcoats . . . in 'foxholes'" (Blunden, 148), and the *Oxford English Dictionary* cites earlier appearances in the *Manchester Guardian* in 1915 and *Red Cross Magazine* in 1919. Comments on the topic, "Foxhole: was the term used in the Great War?" on the website Great War Forum in April 2016, suggest that the term referred to dugouts, cavities dug in trench walls, rather than to a small independent pit for one or two persons as in World War II usage (see "Fox-hole, Great War Forum).

68. Sassoon, *MFM*, 95, 120, 133–34, 145, 226. In *"To Whom the Goddess . . .,"* Shedden and Apsley list wire as first among the six "dangers that beset hunt-ing today. . . . There is no need to more than mention this serious peril which affects all hunting people" (Shedden and Apsley, 271).

69. Sassoon, *MFM,* 273–74. The details of Sassoon's poem, "Wirers" (1917), about a soldier named Hughes killed on a wiring-party, match those of Tiltwood's death in *Fox-Hunting Man* (Sassoon, *The War Poems,* 90).

70. Sassoon, *MFM,* 274–75; *MIO,* 445. Graves reports that this transformation is based directly on Sassoon's experiences: "I felt David's death [David Thomas, Tiltwood's real-life counterpart] worse than any other since I had been in France, but it did not anger me as it did Siegfried. He was acting transport-officer and every evening now, when he came up with the rations, went out on patrol looking for Germans to kill. I just felt empty and lost" (Graves, *Goodbye to All That,* 174).

71. Sassoon, *MFM,* 281–82. Sherston's many references to wire and wire cutting concern the huge tangles of wire that guarded both British and German trenches, obstacles so formidable that heavy tanks, to cite Kenyon again, were developed primarily to remove them: "The job of these vehicles was to crush the wire, to allow the infantry into the German defences, and to destroy machine-gun positions" (Kenyon, *Horsemen in No Man's Land,* 242). Wire strung to impede cavalry was more directly comparable to the wire that challenged Sherston as a foxhunter. DiMarco writes of the Eastern Front, "Wire was a major problem. A few simple strands of wire, not even barbed wire, were sufficient to stop a galloping regiment cold" (DiMarco, *War Horse,* 318). And General John French reported in 1915 on a mounted attack in early fighting on the Western Front, "General De Lisle of the Second Cavalry Brigade . . . formed up and advanced . . . but was held up by wire about five hundred yards from his objective and . . . suffered severely" (French, quoted in Hayne, *Lectures on Cavalry,* 59). Lord Mottistone, with characteristic bravado, recalls, "Through my telescope I had seen that a thin strand of wire ran through the trees, but I reckoned that it would easily be broken. The leading squadron of Dragoons galloped right through it, and on to the ridge" (Mottistone, *My Horse Warrior,* 112).

72. Sassoon, *MIO,* 307, 316, 511. With respect to cavalry in the Boer War, Rankin wrote in *A Subaltern's Letters to His Wife,* "Injuries from barbed wire [to horses] were all too common. . . . A wire-cutter was an instrument on which one's life often depended" (Rankin, 52–53). Badsey notes, "In the Boer War scouts with wirecutters led British cavalry advances, and by the First World War wirecutters were standard equipment" (Badsey, *Doctrine and Reform in the British Cavalry,* 20).

73. Sassoon, *MIO,* 329, 401.

74. Sassoon, *SP,* 640–43. That Sherston "borrows [a] bayonet" for this ultimately redemptive adventure (*SP,* 641) recalls specifically and ironically the "final words" of a training officer's lecture on bayonet use: "There's only one good Boche, and that's a dead one! . . . Stick him between the eyes, in the throat, in the chest" (*MIO,* 290).

75. Rimington, *Our Cavalry,* 19.

76. Tyndale, *A Treatise on Military Equitation,* 8, 60; Maude, *CPF,* 234. DiMarco notes, "British cavalry were notorious for losing control while charging. Even when they were successful . . . they took severe casualties because their com-

manders could not control them. . . . British cavalry [at Waterloo] were eager to gallop into combat—much to Wellington's disgust," though Wellington, one might add, was an avid foxhunter. As the French general Exelmans put it: "The great deficiency is in your officers who seem to be impressed by the conviction that they can dash or ride over everything, as if the art of war were precisely the same as the art of fox-hunting" (Exelmans, quoted in DiMarco, *War Horse*, 215). Littauer observes that England, following the Napoleonic Wars, was a naval power that did not maintain a standing army, so "foxhunting and racing flourished rather than military reviews. . . . The great majority of English 19th century riders then were civilian rather than military, and riding continued to be for sport rather than for parade or the manege" (Littauer, *The Development of Modern Riding*, 137–38).

77. Nolan, *TCRH*, 36; Hayes, *Among Horses in Russia*, 58; Roberts, introduction to Childers, *WAB*, xiv; Rimington, 197. See also, Riedi, "Brains or Polo?" 236–53.

78. The sentiment was not unique to cavalry officers, or, for that matter, to men, military or not. In *"To Whom the Goddess . . . ,"* for example, Shedden and Apsley write, "Many times has England owed much to her young officers trained in the hunting-field, and to the fact that the passion of her leisured classes had been for hunting rather than for luxurious, soft, unhealthy pursuits" (Shedden and Apsey, 36)—a passion, needless to add, shared equally by the bold horsewomen of the leisure classes, such as the authors.

79. Rimington, *Our Cavalry*, 154, 159; Wheeler-Nicholson, *Modern Cavalry*, 49. Not all civilian writers would agree. Much later in the century, for example, the British biographer John Terraine wrote with obvious approval in *Douglas Haig: The Educated Soldier* (1963), "There was very little 'Tally Ho' about [him]" (Terraine, quoted in Kenyon, *Horsemen in No Man's Land*, 4).

80. Rimington, *Our Cavalry*, 158–59, 165, 168; Wheeler-Nicholson, *Modern Cavalry*, 12, 49–50. For a period discussion of equestrian sports in the military during the Great War, see, for example, Clarke, "Mounted Sports," in *Transport and Sport*, 163–71; Dawson, "Hounds Will Meet" and "Front Line Polo," in *Sport in War*, 81–98.

81. Nolan, *CHT*, 61; Wood, *Achievements of Cavalry*, 39; Denison, *MC*, 96; Roosevelt, "Riding to Hounds on Long Island," 341, and see, for recent affirmations of Roosevelt's point, Brereton, *The Horse in War*, 103, and Coulston, introduction to Nolan, *CHT*, xxi.

82. Alderson, *Pink and Scarlet*, vii, 13. A second edition of *Pink and Scarlet*, published in 1913, included twelve mounted color plates by Lionel Edwards, arranged in pairs, across six two-page spreads, with a hunting scene on the left, a corresponding military scene on the right, and a legend spanning the spread.

83. Verney, Lord Willoughby de Broke, *Hunting the Fox*, 1–2; Shedden and Apsley, *"To Whom the Goddess . . . ,"* 40. For a brief period overview of foxhunting in postwar England, see Dixon, "Changed and Changing Conditions—Social and Territorial," the opening chapter of his *Fox-Hunting in the Twentieth Century*, 1–10. Part 1 of this volume, "Changing Conditions," covers general conditions of

238 Notes to Pages 163–168

foxhunting in England during and after the war; part 2, "The History of Ten Years," surveys in varying length "every hunt" in England.

Postscript

1. Brereton, *The Horse in War*, 76, 143.

2. The website for the Land Rover Kentucky Three-Day Event, one of the "grand slam" contests in international eventing competition, explains the connection succinctly under "The History of Eventing": "The tests of this newly organized equestrian competition [in the 1912 Olympic Games] were patterned after the training and testing of military chargers—precision, elegance, and obedience on the parade ground; stamina, versatility and courage on marches and in battle; cross-country jumping ability and endurance in traveling great distances over difficult terrain and formidable obstacles in the relaying of important dispatches; and jumping ability in the arena to prove the horse's fitness to remain in service" (see "History of Eventing").

3. Truscott, *The Twilight of the U. S. Cavalry*, 71, 88; Chamberlin, *Training Hunters Jumpers and Hacks*, x, xv–xvi; Morris, foreword to Matha, *General Chamberlin*, ix. See also Chamberlin, *Breaking, Training and Reclaiming Cavalry Horses*. DiMarco notes "that to Chamberlin and the Army of his time, the goal of equestrian excellence was not about sport [but about] military technical and tactical excellence. . . . Horsemanship did not win cups and trophies, it won battles and wars. Training the horse and rider . . . was about the ability to maneuver on the battlefield . . . to march long distances . . . to carry weight over difficult terrain, and finally . . . to mass hundreds of horses and riders in precise formations, under disciplined control. . . . This was the ultimate goal of Chamberlin's study and practice" (DiMarco, introduction to Matha, *General Chamberlin*, xiv).

4. Bismarck, *Lectures on the Tactics of Cavalry*, 53, 58, 61.

5. As Gustav Wrangel wrote in 1907, "An obedient, well-balanced horse is the first essential for the proper performance of cavalry duty, whether it be patrol work, attacking, or raiding. Therefore the breaking of remounts is, and must without doubt always remain, the most important part of individual training" (Wrangel, *The Cavalry in the Russo-Japanese War*, 58).

6. Renoir's title not only echoes a common French name for the war, "*La Grande Guerre*," but also alludes to Sir Norman Angell's antiwar treatise, *The Great Illusion: A Study of the Relation of Military Power in Nations to Their Economic and Social Advantage* (1909), a widely read work republished in many editions and in many languages between 1909 and 1913 (in French, as *La Grande Illusion*). "The scope of the [study's] whole argument," Angell wrote in the expanded edition of 1912, "is not that war is impossible, but that it is futile—useless, even when completely victorious, as a means of securing those moral or material ends which represent the needs of modern European peoples; and that on a general realization

of this truth depends the solution of the problems of armaments and warfare" (Angell, *The Great Illusion*, v–vi).

7. Desperate to find some *spoken* emotional connection with de Boeldieu, Maréchal cannot bridge the chasm symbolized by de Boeldieu's refusal to *tutoyer* him. Indeed, when Maréchal, exasperated, complains, "I've been with you every day for eighteen months, and you still say *vous* to me," Boeldieu, who continues to comb his hair, stiffly replies, "I say *vous* to my mother and my wife."

8. English as the language that binds aristocrats, not surprisingly, separates other officers, albeit more ambiguously. In a stirring moment, English soldiers in the POW camp lead in the singing of *La Marseillaise,* a moment of communication across languages even if only by rote. But when Maréchal, leaving the first camp, tries to tell an incoming English officer that the departing French have completed an escape tunnel, under the prisoners' quarters and hidden from the Germans, the two cannot understand one another, so the tunnel will remain hidden from the English as well as from the Germans, and never used for escape.

9. Visual symbols of de Boeldieu's and von Rauffenstein's shared class and rank include their monocles and their dress uniforms and gloves, fastidiously maintained even in prison conditions. Visual symbols of their shared horsemanship include the riding crops and spurs that decorate von Rauffenstein's private quarters, a former chapel still dominated by an altar and crucifix, and the photographs of horses that decorate de Boeldieu's meager but well-ordered corner of the prisoners' shared quarters.

10. Over the course of the film, Rosenthal, from a wealthy German-Jewish family, has endured explicit anti-Semitism from both the working-class French Maréchal and the upper-class German von Rauffenstein. Experiencing Rosenthal's generosity of spirit following their escape, and coming to appreciate and understand their solidarity, Maréchal recants. None of the irony would have been lost on an audience in 1937.

11. Renoir, filmed introduction to the restored version of *La Grande Illusion;* Sassoon, *SP,* 557.

Bibliography

Primary Sources

Adams, John. *An Analysis of Horsemanship, Teaching the Whole Art of Riding.* 2 vols. Dedication to the Duke of York, preface, and introduction by Adams. London: Printed by M. Ritchie for the Author, 1799. 2nd ed. 3 vols. London: James Cundee, 1805. "An" was added to the title in 1805.

Alderson, E. A. H. *Pink and Scarlet, or Hunting as a School for Soldiering.* Preface by Alderson. London: William Heinemann, 1900.

Angell, Norman. *The Great Illusion: A Study of the Relation of Military Power to National Advantage.* 1909. London: William Heinemann, 1912. Originally subtitled, *A Study of the Relation of Military Power in Nations to Their Economic and Social Advantage.*

Anonymous [on J. L. Jackson]. "Monthly Catalogue, For September, 1765." *The Monthly Review; or, Literary Journal: By Several Hands.* Vol. 33. London: Printed for R. Griffiths, 1765.

Astley, John. *The Art of Riding, set forth in a breefe treatise [etc].* A Letter Missive to M. Henrie Mackwilliam and M. William Fitzwilliams by Astley. Letter to the Reader by Mackwilliam and Fitzwilliams. London: Henrie Denham, 1584.

Baker, Valentine. *The British Cavalry. With Remarks on Its Practical Organization.* Preface and introduction by Baker. London: Longman, Brown, Green, Longmans, & Roberts, 1858.

Baucher, François. *Dialogues on Equitation.* 1835. Translated with an introduction by Hilda Nelson. In Nelson, ed. *François Baucher: The Man and His Method,* 172–91. Franktown, VA: Xenophon, 2013.

Baucher, François. *A Method of Horsemanship, Founded upon New Principles.* 2nd American ed. Translator unidentified. Philadelphia: A. Hart, 1852.

Bedingfield, Thomas. *The Art of Riding . . . Written at large in the Italian toong, by Maister Claudio Corte.* Epistle to Henrie Mackwilliam and Letter to the Reader by Bedingfield. Epistle to Gentlemen Pensioners by Mackwilliam. London: H. Denham, 1584.

Berenger, Richard. *The History and Art of Horsemanship.* 2 vols. Dedication to the king by Berenger. London: T. Davies, 1771.

Berenger, Richard. *A New System of Horsemanship, From the French of Monsieur Bourgelat.* Translated with a preface by Berenger. London: Printed by Henry Woodfall for Paul Vaillant, 1754.

Bernhardi, Friedrich von. *Cavalry in Future Wars.* Translated by C. S. Goldman. Introduction by Lieut.-General Sir John French. Preface and introduction by Bernhardi. 1906. 2nd ed. Preface to 2nd ed. by Bernhardi added. London: John Murray, 1909. Cited in notes as *CFW.*

Bernhardi, Friedrich von. *Cavalry in War and Peace.* Translated with a translator's note by G. T. M. Bridges. Preface by Field Marshal Sir [John] French. Introduction by Bernhardi. London: Hugh Rees, 1910. Also published as *Cavalry, A Popular Edition of Cavalry in War and Peace.* Edited by A. Hilliard Atteridge, with an editor's note, from the translation of Major G. T. M. Bridges. New York: Doran, 1914. Citations refer to the 1914 edition. Cited in notes as *CWP.*

Bernhardi, Friedrich von. *Germany and the Next War.* Translated by Allen H. Powles. Preface and introduction by Bernhardi. 1912. Authorized American ed. New York: Longmans, Green, 1914. Cited in notes as *GNW.*

Bernhardi, Friedrich von. *How Germany Makes War.* New York: Doran, 1914. Preface by editor. Introduction by Bernhardi. Abridgment of *On War of Today.* 2 vols. Translated by Karl von Donat. London: Hugh Rees, 1912. Cited in notes as *HGMW.*

Bernhardi, Friedrich von. *The War of the Future, In the Light of the Lessons of the World War.* Translated by F. A. Holt. Preface and introduction by Bernhardi. New York: D. Appleton, 1921. Cited in notes as *WF.*

Beudant, Étienne. *Horse Training: Out-door and High School.* Translated with an introduction by John A. Barry. Preface by Th. Monod. 1931. Reissued, New York: Charles Scribner's Sons, 1941.

Bismarck, Friedrich Wilhelm von. *Lectures on the Tactics of Cavalry: And Elements of Manoeuvre for a Cavalry Regiment.* Translated with a preface by N. Ludlow Beamish. London: William H. Ainsworth, 1827. *Lectures* originally published 1818; *Elements of Manoeuvre,* 1819.

Blunden, Edmund. *Undertones of War.* Introduction by Blunden. London: Richard Cobden-Sanderson, 1928.

Blundeville, Thomas. *The Arte of Ryding and Breakinge Greate Horses.* 1560. Facsimile of the 1st ed., New York: Da Capo, 1969.

Blundeville, Thomas. *The Foure Chiefest Offices belonging to Horsemanship.* Dedication to Roberte Dudley and preface by Blundeville. London: Peter Short, 1597.

Boniface, Jonathan. *The Cavalry Horse and His Pack.* Preface by Boniface. 1903. Reprint, Glasgow, KY: The Long Riders' Guild, 2005.

Borden, Spencer. *The Arab Horse.* Preface by Henry Fairfield Osborn. Introduction by Borden. New York: Doubleday Page, 1906.

Borden, Spencer. *What Horse for the Cavalry?* Preface by Borden. Fall River, MA: J. H. Franklin, 1912.

Brittain, Vera. *Testament of Youth.* 1933. Preface by Rt. Hon. Shirley Williams. Foreword by Brittain. Reprint, London: Victor Gollancz, 1979.

Brooke, Geoffrey. *Horse-Sense and Horsemanship of To-Day: Economy and Method in Training Hunters and Polo Ponies.* Introductions by General The Earl of Cavan and Lord Wodehouse. Foreword by Brooke. New York: Charles Scribner's Sons, 1924.

Brooke, Geoffrey. *The Way of a Man with a Horse.* Editors' introduction by the Earl of Lonsdale. Foreword by Brooke. London: Seeley, Service, 1929.

Cairns, D. S. *An Answer to Bernhardi.* Unsigned explanatory note. London: Humphrey Milford, Oxford University Press, 1914.

[Calvert, Sir Harry]. *Regulations and Instructions for Cavalry Sword Exercise.* London: William Clowes, 1819.

Carter, William H. *The American Army.* Author's note by Carter. Indianapolis, IN: Bobbs-Merrill, 1915. Cited in notes as *AA.*

Carter, William H. *Horses, Saddles and Bridles.* 1895. 3rd ed. Preface to 3rd ed. and introduction by Carter. Baltimore, MD: Lord Baltimore, 1906. Cited in notes as *HSB.*

Cavendish, William. 1st Duke of Newcastle. *A General System of Horsemanship, In All It's Branches.* 1743. Facsimile of 1743 ed. with foreword by William C. Steinkraus and technical commentary by E. Schmit-Jensen. London: J. A. Allen, 2000.

Chamberlin, Harry D. *Breaking, Training and Reclaiming Cavalry Horses.* Fort Riley, KS: Second Cavalry Division, 1941.

Chamberlin, Harry D. *Riding and Schooling Horses.* Preface by Chamberlin. Introduction by John Cudahy. New York: Derrydale, 1934.

Chamberlin, Harry D. *Training Hunters Jumpers and Hacks.* Preface by Chamberlin. New York: Derrydale, 1937.

Chenevix-Trench, F. *Cavalry in Modern War.* Preface by Colonel C. B. Brackenbury. London: Kegan Paul, Trench, 1884.

Chichester, Henry Manners. "Lewis Edward Nolan." *Dictionary of National Biography, 1895–1900.* Vol. 41. Accessed March 4, 2020. shttps://en.wikisource. org/wiki/Nolan,_Lewis_Edward_(DNB00)#p-search.

Childers, Erskine. *German Influence on British Cavalry.* Preface by Childers. London: Edward Arnold, 1911. Cited in notes as *GIBC.*

Childers, Erskine. *War and the Arme Blanche.* Introduction by Field-Marshal Earl Roberts. London: Edward Arnold, 1910. Cited in notes as *WAB.*

Cooke, Philip St. George. *Cavalry Tactics or Regulations for the Instruction, Formations, and Movements of the Cavalry of the Army and Volunteers of the United*

States. Preliminary note from Simon Cameron, Secretary of War. 1862. Facsimile of the 1st ed. Mechanicsburg, PA: Stackpole, 2004.

Creasy, Edward S. *Fifteen Decisive Battles of the World*. 1851. Preface by Creasy. Reprint, Harrisburg: Military Service, 1955.

Crowe, J. H. V. *Problems in Manoeuvre Tactics with Solutions for Officers of All Arms*. After the German of Major Hopenstedt. Preface by Crowe. New York: Macmillan, 1905.

Curtiz, Michael, dir. *The Charge of the Light Brigade*. Warner Bros. Pictures, 1936. Burbank, CA: Warner Home Video, 2007. DVD.

Custer, George Armstrong. *My Life on the Plains, or, Personal Experiences with Indians*. 1874. Introduction by Edgar I. Stewart. Reprint, Norman, OK: University of Oklahoma Press, 1962.

D'Aure, Le Comte. Traité d'équitation. Paris: Anselin, Successeur de Magimel, 1834.

Dawson, Lionel. "Hounds Will Meet (War Permitting)" and "Front Line Polo." In *Sport in War*, 83–90, 93–98. New York: Scribner, 1937.

De Broke, Lord Willoughby (Richard Greville Verney). *Hunting the Fox*. Preface by Willoughby de Broke. London: Constable, 1920.

Decarpentry, General. *Baucher and His School*. 1987. Translated by Michael L. M. Fletcher. Foreword by Major Miguel Tavora. Introduction by Richard Williams. Author's foreword by Decarpentry. Reprint, Franktown, VA: Xenophon, 2011.

De la Guérinière, François Robichon. *School of Horsemanship*. 1733. Translated with a preface by Tracy Boucher. Foreword by Paul Belasik. Introduction by Jack C. Schuman. Preface by de la Guérinière. Reprint, London: J. A. Allen, 1994.

Denison, George T. *A History of Cavalry from the Earliest Times, With Lessons for the Future*. 1877. 2nd ed. Prefaces to the 1st and 2nd eds. and introductory chapter by Denison. London: Macmillan, 1913. Cited in notes as *HC*.

Denison, George T. *Modern Cavalry: Its Organisation, Armament, and Employment in War*. Preface and introduction by Denison. London: Thomas Bosworth, 1868. Cited in notes as *MC*.

De Pluvinel, Antoine. *Le Maneige Royal, or, L'Instruction du Roy, En L'Exercise de Monter à Cheval*. 1623. Expanded 1625. Dedication to the king by Pluvinel. Celebratory poems by many hands. Notice to the reader by I. D. Peyrol. Dedication to the king by Chrispian de Pas. Translated with a preface and introduction by Hilda Nelson. Commentary by Elaine Walker. Foreword by William Steinkraus. Edited by Richard F. Williams. Reprint, Franktown, VA: Xenophon, 2015.

Dixon, William Scarth. "Changed and Changing Conditions—Social and Territorial." In *Fox-Hunting in the Twentieth Century*, 1–10. London: Hurst & Blackett, 1925.

Dom Duarte. *The Art of Riding on Every Saddle*. 1434. Translated by António Franco Preto and Luís Franco Preto, with a preface by António Franco Preto (*père*). First published 2006 by Chivalry Bookshelf as *The Royal Book of Jousting, Horsemanship & Knightly Combat.* Reprint, Highland Village, TX: Chivalry Bookshelf Edition, 2011.

Dyer, A. B. *Handbook for Light Artillery.* Preface by Dyer. New York: John Wiley & Sons, 1896.

Eisenberg, Friedrich Wilhelm von. *The Art of Riding a Horse, or, Description of Modern Manège in Its Perfection.* 1727. Dedication to the king and preface by Eisenberg. Translated by Sherilyn Allen. Foreword by Giovanni Battista Tomassini. Edited by Richard F. Williams. Franktown, VA: Xenophon, 2015.

Eltinge, Leroy. "Notes on Cavalry." In *Notes on Infantry, Cavalry, and Field Artillery,* 29–49. Special Reprint for Officers' Training Camps. Washington, DC: Government Printing Office, 1917.

Evans, R. *Outline of the Campaign in Mesopotamia 1914–1918.* Prefatory note by Evans. London: Sifton Praed, 1930.

Faverot de Kerbrech, François. *Methodical Dressage of the Riding Horse* and *Dressage of the Outdoor Horse.* 1891. Edited by Richard and Frances Williams. Translated with a foreword by Michael L. M. Fletcher. Reprint, Franktown, VA: Xenophon, 2010.

Fawcett, William. *Rules and Regulations for the Sword Exercise of the Cavalry.* Cover notice and introduction by Fawcett. London: T. Egerton, 1796. See also, Fawcett, William. *Six Engravings, Representing the Six Cuts in the Sword Exercise of the Cavalry.* Introduction by Fawcett. 8th ed. London: J. Cundee for J. Harris, 1803.

Fillis, James. *Breaking and Riding with Military Commentaries.* Preface by Fillis. Translated with a preface by M. H. Hayes. 1892, 5th ed. London: Hurst and Blackett, 1902.

Findley, Timothy. *The Wars.* New York: Delacorte/Seymour Lawrence, 1977.

Frederick II, King of Prussia. *Military Instructions from the Late King of Prussia to His Generals.* 1762. Translated from the French with dedication and preface to the 1st ed. by Lieut.-Colonel Foster. 5th ed. Sherborne: Printed by and for J. Cruttwell, 1818.

Freeman, Strickland. *The Art of Horsemanship, Altered and Abbreviated, According to the Principles of the Late Sir Sidney Medows.* Dedication to the Prince of Wales and preface by Freeman. London: Printed for the author by W. Bulmer, 1806.

French, E. G. *Good-Bye to Boot and Saddle, or, The Tragic Passing of British Cavalry.* Preface by French. Foreword by General Sir Hubert de la Poer Gough. London: Hutchinson, 1951.

Galtrey, Sidney. *The Horse and the War.* Note by Field Marshal Douglas Haig. Introductory chapter by Galtry. London: Country Life, 1918.

Gambado, Geoffrey. *An Academy for Grown Horsemen, Containing the Complete Instructions for Walking, Trotting, Cantering, Galloping, Stumbling, and Tumbling* (1787), and *Annals of Horsemanship: Containing Accounts of Accidental*

Experiments, and Experimental Accidents Both Successful and Unsuccessful (1791). Published in multiple contemporary editions separately and as two volumes in one, the latter, for instance, London: Printed for Hooper & Wigstead, 1796.

Garrard, Kenner. *Nolan's System for Training Cavalry Horses.* Preface by Garrard. New York: D. Van Nostrand, 1862.

General Service Schools. *Tactics and Technique of Cavalry, 1921.* Introduction by H. E. Ely, commandant. Leavenworth, KS: General Services Schools, 1922. See also, *Tactics and Technique of Cavalry: A Text and Reference Book of Cavalry Training.* 1930. 6th ed. Washington, DC: Military Service, 1935.

General Staff. *Handbook of the German Army.* 1917. Translator unidentified. Reprint, London: EP Publishing, 1973.

Gilbey, Walter. *The Great Horse, or The War Horse.* 1889. 2nd ed. Preface to 2nd ed. and introduction by Gilbey. London: Vinton, 1899.

Gilbey, Walter. *Horse-Breeding in England and India and Army Horses Abroad.* 1884. 2nd ed. Preface by Gilbey. London: Vinton, 1906.

Gilbey, Walter. *Horses for the Army: A Suggestion.* 1902. Preliminary note by Gilbey, "A Suggestion," added to 1913 ed. London: Vinton, 1913.

Gilbey, Walter. *Small Horses in Warfare.* Preliminary note by Gilbey. London: Vinton, 1900.

Goldman [misspelled Goldmann], Charles Sidney. *With General French and the Cavalry in South Africa.* Preface and introductory chapter by Goldman. London: Macmillan, 1902.

Government of India. *Report of the Horse and Mule Breeding Commission.* Umballa: India Office, 1901.

Graves, Robert. *Goodbye to All That.* 1929. New ed., revised. Prologue by Graves. London: Cassell, 1957.

Gray, Alonzo. *Cavalry Tactics as Illustrated by the War of the Rebellion.* Preface and introduction by Gray. Fort Leavenworth, KS: U.S. Cavalry Association, 1910.

Grisone, Federico. *The Rules of Riding.* 1550. Edited with an introduction by Elizabeth MacKenzie Tobey. Translated by Tobey and Federica Brunori Deigan. Preface by David Guy. Tempe, AZ: Arizona Center for Medieval and Renaissance Studies, 2014.

Haig, Sir Douglas. *Cavalry Studies: Strategical and Tactical.* Preface and introductory chapter by Haig. London: Hugh Rees, 1907.

Haking, R. C. B. *Staff Rides and Regimental Tours.* Preface by Haking. London: Hugh Rees, 1908.

Hayes, M. Horace. *Among Horses in Russia.* Preface by Hayes. London: R. A. Everett, 1900.

Hayne, P. T. *Lectures on Cavalry.* Leavenworth: Army Service Schools, 1915.

H.Dv.12: Army Riding Regulations [German Cavalry Manual on the Training of Horse and Rider]. 1937. Preface to the English ed. by Richard F. Williams.

Translated with an introduction by Stefanie Reinhold. Forewords by Eckart Meyners and Christoph Hess. Reprint, Franktown, VA: Xenophon, 2014.

Hemingway, Ernest. *Death in the Afternoon.* Explanatory glossary and bibliographical note by Hemingway. New York: Charles Scribner's Sons. 1932.

Hemingway, Ernest. *A Farewell to Arms.* New York: Charles Scribner's Sons. 1929.

Hemingway, Ernest. *In Our Time.* 1924, 1925, 1930. Introductions by Edmund Wilson and Hemingway. Reprint, New York: Charles Scribner's Sons, 1931.

Henderson, Ernest F. *Germany's Fighting Machine.* Indianapolis: Bobbs-Merrill, 1914.

Herbert, Henry, 10th Earl of Pembroke. *Military Equitation, or, A Method of Breaking Horses, and Teaching Soldiers to Ride.* Dedication to the king by Pembroke. London: J. Hughes, 1761. 4th ed. London and Salisbury: Printed for G. and T. Wilkie and E. and J. Easton, 1793.

Hershberger, H. R. *A Sabre Exercise for Mounted and Dismounted Service*, 107–41. In *The Horseman, A Work on Horsemanship,* by Hershberger. Preface by Hershberger. New York: Henry G. Langley, 1844.

Hinde, Captain [Robert]. *The Discipline of the Light-Horse.* Dedication to Lieutenant-General Carpenter and Compliments to the Officers of Light-Dragoons by Hinde. London: Printed for R. Owen, 1778.

Hinds, John [John Badcock]. *Rules for Bad Horsemen: Hints to Inexpert Travellers; and Maxims worth remembering by the most Experienced Equestrians. A New Edition, with Modern Additions.* London: Published for the author by Sherwood and Co, 1830.

Hughes, Charles. *The Compleat Horseman; or, The Art of Riding Made Easy.* Dedication by Hughes. London: Printed for F. Newbery, 1772.

Huidekoper, Frederic Louis. *The Military Unpreparedness of the United States.* Preface by Huidekoper. Introduction by Major General Leonard Wood. New York: Macmillan, 1915.

Huth, F. H. *Works on Horses and Equitation: A Bibliographical Record of Hippology.* Preface by Huth. 1887. Reprint, Hildesheim, Germany: Olms, 1981.

Imperial War Museum. *A Concise Catalogue of Paintings, Drawings and Sculpture of the First World War 1914–1918.* 1924. 2nd ed. Introduction to 1st ed. by Martin Conway. Preface by Noble Frankland. London: Imperial War Museum, 1963.

Ingelfingen, Prince Kraft zu Hohenlohe. *Letters on Cavalry.* Translated with a preface by N. L. Walford. London: Edward Stanford, 1889.

Jackson, J. L. *The Art of Riding; or, Horsemanship made Easy.* Preface by Jackson. London: Printed for A. Cooke, 1765.

Johnston, James Robert. *Riding into War: The Memoir of a Horse Transport Driver, 1916–1919.* Introduction by Brent Wilson. Fredericton, NB, Canada: Goose Lane, 2004.

Lawrence, John. *A Philosophical and Practical Treatise on Horses and on the Moral Duties of Man towards the Brute Creation.* Preface and introductory chapter by

Lawrence. 1796. 2nd ed. 2 vols. London: C. Whittingham, for H. D. Symonds, 1802. Citations refer to the 1802 ed.

L'Hotte, Alexis François. *Un Officier de cavalerie, Souvenirs.* Unsigned editorial preface. Paris: Librairie Plon, 1905.

L'Hotte, Alexis-François. *Questions équestres.* 1895. Preface to 1906 ed. by anonymous editor, H. N. Translated by Hilda Nelson. In *Alexis-François L'Hotte: The Quest for Lightness in Equitation*, edited by Nelson, 147–214. London: J. A. Allen, 1997.

Liddell Hart, B. H. "After Cavalry—What?" *Atlantic Monthly* 136 (July–December 1925): 409–18. Cited in notes as "ACW."

Liddell Hart, B. H. *The Remaking of Modern Armies.* Preface by Liddell Hart. 1927. Boston: Little, Brown, 1928. Cited in notes as *RMA.*

Liebknecht, Karl. *Militarism.* Translator unidentified. Introduction by Sidney Zimand. New York: B. W. Huebsch, 1917.

Littauer, Vladimir. *Russian Hussar.* Foreword by Sir Robert Bruce Lockhart. London: J. A. Allen, 1965.

Loir, Captain. *Cavalry. Technical Operations. Cavalry in an Army. Cavalry in Battle.* Translated by General Staff, War Office. Preface by General Langlois. London: Harrison and Sons, 1916.

Mach, Edmund von. *What Germany Wants.* Personal foreword by Mach. Boston: Little, Brown, 1914.

Markham, Gervase. *Cheape and Good Husbandry for The Well-Ordering of all Beasts and Fowles, and for the generall Cure of their Diseases.* Epistle Dedicatory to Richard Sackville, and To the Curious Reader, both by Markham. London: Printed by T. S. for Roger Jackson, 1614. 8th ed., London: Printed by Thomas Harper, 1653. Citations refer to the 1653 ed.

Maude, F. N. *Cavalry: Its Past and Future.* Preface by Maude. London: William Clowes & Sons, 1903. Cited in notes as *CPF.*

Maude, F. N. *Cavalry Versus Infantry.* Editor's preface by Captain Arthur L. Wagner. Kansas City, MO: Hudson-Kimberly, 1896. Cited in notes as *CVI.*

Maydon, J. G. *French's Cavalry Campaign.* Preface and introduction by Maydon. London: C. Arthur Pearson, 1901.

McClellan, George. *European Cavalry, Including Details of the Organization of the Cavalry Service among the Principal Nations of Europe.* Unsigned publisher's preface. Philadelphia: J. B. Lippincott, 1861.

Monsenergue, Colonel. *Cavalry Tactical Schemes: A Series of Practical Exercises for Cavalry.* Translated by E. Louis Spiers. Introduction by Brig.-General H. de la P. Gough. Preface by F. Bennett-Goldney. London: Hugh Rees, 1914.

Morgan, Nicholas. *The Perfection of Horsemanship, drawne from Nature, Arte, and Practise.* Epistles to the king, Prince Henry, Edward Earle of Worcester, Gentlemen of great Brittaine, and Poem to Robert Alexander, all by Morgan. London: Imprinted at London for Edward White, 1609.

Mottistone, Lord (General Jack Seely). *My Horse Warrior.* Foreword by Mottistone. 1934. London: Hodder & Stoughton, 1938.

Neville, Capt. L. *A Treatise on the Discipline of Light Cavalry.* Introductory note by Neville. London: Printed for T. Egerton, 1796.

Nolan, Lewis Edward. *Cavalry: Its History and Tactics.* 1853. Preface by Nolan. Introduction by Jon Coulston. Reprint, Yardley, PA: Westholme, 2007. Cited in notes as *CHT.*

Nolan, Lewis Edward. *The Training of Cavalry Remount Horses, A New System.* Note and introduction by Nolan. Letters of endorsement by Lieut.-General John Aitchison, M.-General Lovell B. Lovell, Lieut.-Colonel G. W. Key. London: Parker, Furnivall & Parker, 1852. Cited in notes as *TCRH.*

Norman, W. W. *Cavalry Reconnaissance.* Preface and introduction by Norman. London: Hugh Rees, 1911.

O'Reilly, Laurence. *The Art of Horsemanship.* Dedication and preface by O'Reilly. Newry, Ireland: Printed by Dan. Carpenter, 1780.

O'Sullivan, Thaddeus. *Into the Storm.* New York: Home Box Office. Broadcast on HBO, May 31, 2009.

Parker, R. M. *An Officer's Notes.* 4th ed. Compiled with a preface by C. C. Griffith. Letter of endorsement by Major-General Leonard Wood. New York: George U. Harvey, 1917.

Pelet-Narbonne, General von. *Cavalry on Service, Illustrated by the Advance of the German Cavalry Across the Mosel in 1870.* Translated with a preface by Major D'A. Legard. Preface by Pelet-Narbonne. London: Hugh Rees, 1906.

Preston, R. M. P. *The Desert Mounted Corps, An Account of the Cavalry Operations in Palestine and Syria, 1917–1918.* Introduction by Lieut.-General Sir H. G. Chauvel. London: Constable, 1921. Reprinted as *War in Syria.* Edited and introduced by Paul Rich. Washington, DC: Westphalia, 2013. Citations refer to the 2013 ed.

Rankin, Reginald. *A Subaltern's Letters to His Wife.* 1901. 2nd ed. 7th impression. Preface to 2nd ed. by Rankin. London: Longmans, Green, 1901.

Reese, H. H. "Breeding Horses for the United States Army," 341–56. In *Yearbook of the United States Department of Agriculture.* Washington, DC: Government Printing Office, 1918.

Renoir, Jean, dir. *La Grande Illusion. Realisations d'Art Cinematographique,* 1937; World Pictures Corporation, 1938. Criterion Collection, 1999. DVD. Filmed introduction by Renoir, in 1958, to the restored version of the film. YouTube, May 24, 2009. Accessed September 29, 2018. www.youtube.com/watch?v=163xhUUZOH8 See also *Grand Illusion, a film by Jean Renoir.* Translated by Marianne Alexandre and Andrew Sinclair. Classic Film Scripts. New York: Simon and Schuster, 1968.

Richardson, Tony, dir. *The Charge of the Light Brigade.* MGM, 1968. MGM Home Entertainment, 2002. DVD.

Rimington, M. F. *Our Cavalry.* Preface and introductory chapter by Rimington. London: Macmillan, 1912.

Rodzianko, Paul. *Modern Horsemanship.* London: Seeley Service, 1936.

Rodzianko, Paul. *Tattered Banners: An Autobiography.* 1939. Foreword by Gary Saul Morson. Reprint, Philadelphia, PA: Paul Dry, 2018.

Rommel, George M. *The Army Remount Problem.* Washington, DC: Government Printing Office, 1911. Reprinted from the 27th Annual Report of the Bureau of Animal Industry (1910).

Roosevelt, Theodore. "Riding to Hounds on Long Island" in "Cross-Country Riding in America." *Century Illustrated Magazine* 32, no. 3 (July 1886): 335–42.

Russell, Capt. J. C. *Notes on Cavalry Service.* Preface and introductory chapter by Russell. London: Cassell, Petter, & Galpin, 1873.

Santini, Piero. *The Caprilli Papers.* Translated and edited by Santini. Biographical foreword by Lieut.-Col. C. E. G. Hope. Introduction by Santini. London: J. A. Allen, 1967.

Santini, Piero. *The Forward Impulse.* Preface by Santini. New York: Huntington, 1936.

Santini, Piero. *Learning to Ride.* New York: Greenberg, 1941.

Santini, Piero. *Riding Reflections.* Foreword by Lida L. Fleitmann. New York: Derrydale, 1932.

Sassoon, Siegfried. *Diaries 1915–1918.* Edited with an introduction by Rupert Hart-Davis. London: Faber and Faber, 1983.

Sassoon, Siegfried. *The Memoirs of George Sherston.* London: Faber and Faber, 1937. Reprinted as *The Complete Memoirs of George Sherston.* 4th printing. Faber and Faber, 1964. Reprint comprises *Memoirs of a Fox-Hunting Man*, 7–282; *Memoirs of an Infantry Officer*, 283–514; *Sherston's Progress*, 515–656. Page references are to the 1964 ed. and are cited in notes as *MFM, MIO, SP.*

Sassoon, Siegfried. *Siegfried's Journey.* 1945. Reprint, London: Faber & Faber, 1947. Cited in notes as *SJ.*

Sassoon, Siegfried. *The War Poems of Siegfried Sassoon.* Arranged with an introduction by Rupert Hart-Davis. London: Faber & Faber, 1983.

Saxe, Field-Marshal Count de. *Reveries, or Memoirs upon the Art of War.* Translator unidentified. Dedication to the general officers by translator. Preface by Saxe. London: Printed for J. Nourse, 1757.

Schmidt, Carl von. *Instructions for the Training, Employment, and Leading of Cavalry.* 1881. Compiled by Captain von Vollard-Bockelberg. Translated with an introductory note by C. W. Bowdler Bell. Reprint, New York: Greenwood, 1968.

Seeger, Louis. [Excerpts from] *M. Baucher and His Art: A Serious Word with the Riders of Germany.* 1853. Translated by Michael L. M. Fletcher. In *Baucher and His School,* by General Decarpentry, 115–59. Franktown, VA: Xenophon, 2011.

Shedden, Lady Diana, and Lady Apsley. *"To Whom the Goddess . . .": Hunting and Riding for Women.* Introduction by the Earl of Lonsdale. 5th impression. London: Hutchinson, 1932.

Shirley, Arthur. *Remarks on the Transport of Cavalry and Artillery.* London: Parker, Furnivall, and Parker, 1854.

Solleysel, Jacques de. *Le parfait Mareschal* (1664). Translated from the "eighth edition of the original" by Sir William Hope as *The Complete Horseman Discovering the Surest Marks of the Beauty, Goodness, Faults and Imperfections of Horses.* Two parts in 1 vol. London: Printed for M. Gillyflower, et al., 1696. Dedicatory epistle to the king, preface, and author's epistle to the reader by Hope. Unsigned advertisement by the publisher. Abridged ed., London: Printed for H. Benwicke, 1702.

Spaulding, Oliver. *Notes on Field Artillery for Officers of All Arms.* 1908. 4th ed. Prefaces to 1st–4th eds. by Spaulding. Leavenworth, KS: U.S. Cavalry Association, 1918.

Steinbrecht, Gustav. *The Gymnasium of the Horse.* 1886. Translated by Helen K. Gibble. Foreword by William Steinkraus. Preface for reprint by Hans Heinrich Brinckmann. Preface by publisher of 1st ed. by Paul Plinzner. Introduction to 4th ed. by Hans von Heydebreck. Reprint, Franktown, VA: Xenophon, 1995.

Stone, Oliver, dir. *Platoon.* Hemdale Film Corporation/Orion Pictures, 1986. DVD.

Tamblyn, D. S. *The Horse in War: Horses & Mules in the Allied Armies during the First World War, 1914–18.* Originally published as *"The Horse in War"* and *Famous Canadian War Horses.* 1932. Foreword by A. W. Currie. Reprint, Kingston, Ontario: Jackson, n.d.

Temple Clarke, A. O. *Transport and Sport in the Great War Period.* London: Quality, 1938.

Thompson, Charles. *Rules for Bad Horsemen, Addressed to the Society for the Encouragement of Arts, Etc.* Dublin: Printed by James Hoey, 1762. Also 3rd ed., London: J. Robson, 1765; 4th ed., London: J. Robson, 1775. See also, Hinds, John [John Badcock]. *Rules for Bad Horsemen: Hints to Inexpert Travellers; and Maxims worth remembering by the most Experienced Equestrians. A New Edition, with Modern Additions.* 1830.

Truscott, Lucian. *The Twilight of the U. S. Cavalry: Life in the Old Army, 1917–1942.* Edited with a preface by Lucian K. Truscott III. Foreword by Edward M. Coffman. Lawrence: University Press of Kansas, 1989.

Tyndale, W. *Instructions for Young Dragoon Officers.* 2nd ed. Dedication to Earl of Harrington by Tyndale. London: Printed for T. Egerton, 1796.

Tyndale, W. *A Treatise on Military Equitation.* London: Printed for the author by T. Egerton, 1797.

Urban, Sylvanus [on Sidney Medows]. *The Gentleman's Magazine: and Historical Chronicle.* For the year 1792. Vol. 62. Part 2. London: John Nichols, 1792.

Wagner, Arthur L. *Organization and Tactics.* 1894. 5th ed. Prefaces to the 2nd and 1st eds. and introductory chapter, all by Wagner. Kansas City, MO: Hudson-Kimberly, c. 1905. Cited in notes as *OT.*

Wagner, Arthur L. *The Service of Security and Information.* 1893. 11th ed. Prefaces to the 9th, 3rd, and 1st eds. and introductory chapter, all by Wagner. Kansas City, MO: Hudson-Kimberly, c. 1905. Cited in notes as *SSI.*

War Department. *Notes on Equitation and Horse Training.* 1910. Introductory statement by Wm. H. Carter. Facsimile of 1st ed. Provo, UT: Triton, 1988.

Warnery, Emanuel von. *Remarks on Cavalry.* Translated with a dedication to the Duke of York by G. F. Koehler. Preface and introduction by Warnery. 1798. Introduction by Brent Nosworthy. Reprint, London: Constable 1997.

Wells, H. G. *War and the Future: Italy, France and Britain at War.* London: Cassell, 1917.

Weyrother, Max Ritter von. *Fragments from the Writings.* 1836. Translated with translator's note by H. J. Fane. Preface by Andreas Hausberger. Introduction by Daniel Pevsner. Unsigned foreword by Weyrother's students. Reprint, Franktown, VA: Xenophon, 2017.

Wheeler-Nicholson, Malcolm. *Modern Cavalry: Studies on Its Rôle in the Warfare of To-Day with Notes on Training for War Service.* Preface by Wheeler-Nicholson. New York: Macmillan, 1922.

Whyte-Melville, G. J. *Riding Recollections.* Preface by Whyte-Melville. 1878. Reprint, London: Ward, Lock, n.d.

Wingfield, W. J. R. *Lectures to Cavalry Subalterns of the New Armies.* Preface by Wingfield. London: Forster Groom, 1915.

Wood, Sir Evelyn. *Achievements of Cavalry.* Preface by Wood. London: George Bell & Sons, 1897.

Woodhull, Maxwell. *West Point in Our Next War: The Only Way to Create and Maintain an Army.* Introduction by Woodhull. New York: G. P. Putnam's Sons, 1915.

Wrangel, Gustav. *The Cavalry in the Russo-Japanese War: Lessons and Critical Considerations.* Translated with a preface by J. Montgomery. Introduction by Wrangel. London: Hugh Rees, 1907.

Xenophon. *The Art of Horsemanship.* Translated with a preface by Morris H. Morgan, 1894. Reprint, London: J. A. Allen, 1962.

Xenophon. *The Cavalry General.* In *The Works of Xenophon,* n.p. Translated by H. G. Dakyns. London: Macmillan, 1890.

Secondary Sources

Anderson, Dorothy. "Baker, Valentine [*called* Baker Pasha]." September 23, 2004. *Oxford Dictionary of National Biography.* https://doi.org/10.1093/ref:odnb/1142.

Anglesley, Marquess of. *A History of the British Cavalry.* 8 vols. London: Leo Cooper, 1973–1997.

"Animals in War Memorial." The Royal Parks. Accessed May 2, 2020. https://www.royalparks.org.uk/parks/hyde-park/things-to-see-and-do/memorials,-fountains-and-statues/animals-in-war-memorial.

"Animals in War Memorial—The Monument." The Animals in War Memorial. Accessed May 2, 2020. http://www.animalsinwar.org.uk/.

"Ascot Unveils War Horse Memorial to Commemorate WW1." BBC News. 8 June 2018. https://www.bbc.com/news/uk-england-berkshire-44410854.

Badsey, Stephen. *Doctrine and Reform in the British Cavalry 1880–1918.* Burlington, VT: Ashgate, 2008.

Baker, Donald G. "The Only Mounted Unit in the U. S. Army—The 287th M. P. Horse Platoon [in Germany]." *Chronicle of the Horse* 11, no. 1 (November 1, 1957): 13–14.

Brereton, J. M. *The Horse in War.* London: David & Charles, 1976.

Brereton, T. R. *Educating the U. S. Army, Arthur L. Wagner and Reform, 1875–1905.* Lincoln, NE: University of Nebraska Press, 2000.

Brown, David Alan. Introduction to *Virtue and Beauty,* by Brown et al. Published in conjunction with *Leonardo's Ginevra de' Benci and Renaissance Portraits of Women,* organized by and presented at the National Gallery of Art September 30, 2001–January 6, 2002. Washington, DC: National Gallery of Art, 2001.

Butler, Simon. *The War Horses: The Tragic Fate of a Million Horses in the First World War.* Foreword by General Sir Frank Kitson. Wellington, Somerset, UK: Halsgrove, 2011.

Caramello, Charles. "Revisiting Piero Santini, Apostle of Forward Riding (Parts I and II)." *Eventing Nation.* January 20 and 21, 2018. www.eventingnation.com.

Cayce, Gale. "Fox Hunting, Past and Present." Unpublished lecture courtesy of author.

Chenevix-Trench, Charles. *A History of Horsemanship.* Garden City: Doubleday, 1970.

Christie's [Nicholas Morgan]. *Sale 7300.* London. January 26–27, 2006.

Clarke, A. O. Temple. *Transport and Sport in the Great War Period.* London: Quality, 1938.

Coulston, Jon. Introduction to *Cavalry: Its History and Tactics,* by Lewis Edward Nolan. 1853. xi–xxxiv. Reprint, Yardley, PA: Westholme 2007.

Courtney, W. P. "Berenger, Richard." (1885). Revised by S. J. Skedd, September 23, 2004. *Oxford Dictionary of National Biography.* https://doi.org/10.1093/ref:odnb/2193.

Dawson, Anthony. *Real War Horses: The Experiences of the British Cavalry 1814–1914.* Barnsley, South Yorkshire, UK: Pen & Sword Military, 2016.

Dawson, Lionel. *Sport in War.* New York: Charles Scribner's Sons, 1937.

De Bragança, Dom Diogo. *Dressage in the French Tradition.* Translated from Portuguese to French by René Bacharach, with prefaces by de Bragança and by Bacharach, as *L'équitation de Tradition Français,* 2005. Translated from

French to English by Michael L. M. Fletcher, with introduction by Jean Philippe Giacomini. Franktown, VA: Xenophon, 2011.

DiMarco, Louis A. *War Horse: A History of the Military Horse and Rider.* Yardley, PA: Westholme, 2008.

Donovan, Tom. *In Memoriam: A Bibliography of the Personal Memorial Volumes of the Great War, 1914–1918.* Brighton, UK: Tom Donovan Editions, 2015.

Dorondo, David R. *Riders of the Apocalypse: German Cavalry and Modern Warfare 1870–1945.* Annapolis, MD: Naval Institute, 2011.

Drabble, Margaret, ed. with multiple advisors and contributors. *The Oxford Companion to English Literature.* 1932. 6th ed., 2000, revised 2006. New York: Oxford University Press, 2006.

Edwards, Peter. *Horses and the Aristocratic Lifestyle in Early Modern England: William Cavendish, First Earl of Devonshire (1551–1626), and his Horses.* Woodbridge, Suffolk, UK: Boydell, 2018.

"8 Memorials to Animals in the First World War." Heritage Calling. Accessed May 2, 2020. https://heritagecalling.com/2017/02/28/8-memorials-to-animals-in-the-first-world-war/.

"End of Glory: 'Into the Storm.'" International Churchill Society. Accessed May 2, 2020. https://winstonchurchill.org/resources/in-the-media/churchill-in-the-news/end-of-glory-into-the-storm/.

Fairley, John. *Horses of the Great War: The Story in Art.* Barnsley, South Yorkshire, UK: Pen & Sword Military, 2016.

Felton, W. Sidney. *Masters of Equitation.* Foreword by Henry Wynmalen. London: J. A. Allen, 1962.

"Foxhole: was the term used in the Great War?" Great War Forum. April 2016. Accessed December 11, 2018. https://www.greatwarforum.org/topic/238542-foxhole-was-the-term-used-in-the-great-war/?page=3.

Fussell, Paul. *The Great War and Modern Memory.* New York: Oxford University Press, 1975.

Grimshaw, Anne. *The Horse: A Bibliography of British Books, 1851–1976.* Foreword by Dorian Williams. London: Library Association, 1982.

Guy, David. Preface to *The Rules of Riding,* by Federico Grisone, xv–xxii. Edited with an introduction by Elizabeth MacKenzie Tobey. Translated by Tobey and Federica Brunori Deigan. Tempe, AZ: Arizona Center for Medieval and Renaissance Studies, 2014.

Henriquet, Michel. *30 Years with Master Nuno Oliveira.* Franktown, VA: Xenophon, 2011.

"History of Eventing." Land Rover Kentucky Three-Day Event. https://kentuckythreedayevent.com/about-eventing/. Accessed December 15, 2020.

Hochschild, Adam. *To End All Wars: A Story of Loyalty and Rebellion, 1914–1918.* Boston: Houghton Mifflin Harcourt, 2011.

Holmes, Richard. *Tommy: The British Soldier on the Western Front, 1914–1918.* London: HarperCollins, 2004.

Huggins, Mike. *Horse Racing and British Society in the Long Eighteenth Century.* Woodbridge, Suffolk, UK: Boydell, 2018.

Hyland, Ann. *The Warhorse in the Modern Era: The Boer War to the Beginning of the Second Millennium.* Foreword and introduction by Hyland. Stockton-on-Tees, UK: Black Tent, 2010.

Jastrzembski, Frank. "The Reinvention of Valentine Baker." Military History Now. May 20, 2017. https://militaryhistorynow.com/2017/03/20/the-reinvention -of-valentine-baker-how-a-disgraced-british-army-officer-salvaged-his-career -in-a-foreign-war/.

Jones, Spencer. *From Boer War to Great War: Tactical Reform of the British Army, 1902–1914.* Norman: University of Oklahoma Press, 2013.

Kenyon, David. *Horsemen in No Man's Land: British Cavalry and Trench Warfare, 1914–1918.* Foreword by Richard Holmes. Barnsley, South Yorkshire, UK: Pen & Sword Military, 2012.

Lengel, Edward. *World War 1 Memories: An Annotated Bibliography of Personal Accounts Published in English since 1919.* Lanham, MD: Scarecrow, 2004.

Lijsen, H. J. *Classical Circus Equitation: Liberty, High School, Quadrilles and Vaulting.* Translated by Antony Hippisley Coxe. 1949. Reprint, London: J. A. Allen, 1993.

Littauer, Vladimir. *The Development of Modern Riding.* Foreword by William C. Steinkraus. Preface to 1st ed. by Littauer. Originally published as *Horseman's Progress,* 1962. Reprint, New York: Macmillan, 1991.

Livingston, Phil, and Ed Roberts. *War Horse: Mounting the Cavalry with America's Finest Horses.* Preface by Raymond G. Smith. Albany, TX: Bright Sky, 2003.

Macdonald, Janet. *Horses in the British Army 1750–1950.* Barnsley, South Yorkshire, UK: Pen & Sword Military, 2017.

Matha, Warren C. *General Chamberlin: America's Equestrian Genius.* With forewords by George H. Morris and James Wofford, introduction by Louis A. DiMarco, and author's note by Matha. Franktown, VA: Xenophon, 2019.

Moore, Lucinda. *Animals of the Great War: Rare Photographs from Wartime Archives.* Barnsley, South Yorkshire, UK: Pen & Sword Military, 2017.

Moyse-Bartlett, H. *Nolan of Balaclava and His Influence on the British Cavalry.* Originally published as *Louis Edward Nolan,* 1971. Reprint, London: Leo Cooper, 1975.

Nelson, Hilda. *Alexis-François L'Hotte: The Quest for Lightness in Equitation.* Foreword by Colonel François de Beauregard. London: J. A. Allen, 1997.

Nelson, Hilda. *The Ecuyere of the 19th Century in the Circus, with an epilogue on four contemporary écuyeres.* Foreword by Dominique Jando. Franktown, VA: Xenophon, 2001.

Nelson, Hilda. *François Baucher: The Man and His Method.* Publisher's preface by Richard and Frances Williams. Forwards to current ed. by Paul Belasik and to 1st ed. by Charles Harris. London: J. A. Allen, 1992. Reprint, Franktown, VA: Xenophon, 2013.

"100 Years On, USA's Million War Horses Honored." *Horse Talk*, March 11, 2017. https://www.horsetalk.co.nz/2017/03/11/usa-million-war-horses-honored-ww1/.

Ottevaere, James A. *American Military Horsemanship: The Military Riding Seat of the United States Cavalry, 1792 through 1944.* Bloomington, IN: Author House, 2005.

Piekalkiewicz, Janusz. *The Cavalry of WWII.* 1976. New York: Stein and Day, 1980.

Podeschi, John. *Books on the Horse and Horsemanship, 1400–1941.* Compiled with an introduction by Podeschi. London: Tate Gallery for the Yale Center for British Art, 1981.

Poggioli, Renato. *The Theory of the Avant-Garde.* Translated by Gerald Fitzgerald. Originally published in Italian in 1962. Cambridge, MA: Harvard University Press, 1968.

Poscharnigg, Werner. *Austrian Art of Riding: Five Centuries.* Forewords by Karl Mikolka, Charles de Kunffy, Sylvia Loch. Franktown, VA: Xenophon, 2015.

Pyhrr, Stuart W. et al. *The Armored Horse in Europe 1480–1620.* Published in conjunction with the exhibition "The Armored Horse in Europe, 1480–1620," held at the Metropolitan Museum of Art, February 15, 2005 to January 15, 2006. New Haven, CT: Metropolitan Museum/Yale University Press, 2005.

Racinet, Jean-Claude. *Racinet Explains Baucher.* Franktown, VA: Xenophon, 1997.

Riedi, Eliza. "Brains or Polo? Equestrian Sport, Army Reform and the Gentlemanly Officer Tradition, 1900–1914." *Journal of the Society for Army Historical Research* 84, no. 339 (Autumn 2006): 236–53.

R. L. B. "The British Army's Last Mules." *Chronicle of the Horse* 6, no. 20 (June 20, 1975): 59.

Schmit-Jensen, E. Technical commentary on *A General System of Horsemanship, In All It's Branches,* by William Cavendish. London: J. A. Allen, 2000.

Schuessler, Raymond. "The Horse of World War I." *Chronicle of the Horse* 2, no. 23 (February 23, 1973): 24–25.

Screen, J. E. O. "Herbert, Henry, tenth earl of Pembroke." May 21, 2009. *Oxford Dictionary of National Biography.* https://doi.org/10.1093/ref:odnb/13034.

Steggle, Matthew. "Markham, Gervase." September 28, 2006. *Oxford Dictionary of National Biography.* https://doi.org/10.1093/ref:odnb/18065.

Sweetman, John. *Cavalry of the Clouds: Air War over Europe 1914–1918.* Stroud, Gloucestershire, UK: History, 2010.

Tobey, Elizabeth MacKenzie. Introduction to *The Rules of Riding,* by Federico Grisone. 1–57. Edited by Tobey. Translated by Tobey and Federica Brunori Deigan. Tempe, AZ: Arizona Center for Medieval and Renaissance Studies, 2014.

Tomassini, Giovanni. *The Italian Tradition of Equestrian Art: A Survey of the Treatises on Horsemanship from the Renaissance and the Centuries Following.* Translated by author, edited by Richard F. Williams. Forewords by Arthur Kottas-Heldenberg and Joáo Pedro Rodrigues. Franktown, VA: Xenophon, 2014.

Travis, Lorraine. *The Mule*. London: J. A. Allen, 1990.

Van der Horst, Koert, ed. *Great Books on Horsemanship: The Library of Johan Dejager*. Introduction by Johan Dejager. Leiden, Netherlands: Brill, 2014.

W. F. B. "The 287th M. P. Horse Platoon—In Memorium." *Chronicle of the Horse* 23, no. 20 (January 15, 1960): 20.

Wikipedia. "Horse Memorial." Last edited June 11, 2020. https://en.wikipedia.org/wiki/Horse_Memorial.

Winton, Graham. *"Theirs Not to Reason Why": Horsing the British Army 1875–1925*. Foreword by Stephen Badsey. Solihull, West Midlands, UK: Helion, 2013.

Wohl, Robert. *The Generation of 1914*. Cambridge: Harvard University Press, 1979.

Wortley, Laura. *Lucy Kemp Welch, 1869–1958: The Spirit of the Horse*. Woodbridge, Suffolk, UK: Antique Collectors' Club, 1996.

Wrangel, Alexis. *The End of Chivalry: The Last Great Cavalry Battles, 1914–1918*. Foreword by Major-General James D. Lunt. Prefatory note by General John Waters. New York: Hipporene, 1982.

Index

Academie d'Equitation (Manège des Tuileries), 23, 24

Adams, John, 73–74; *An Analysis of Horsemanship*, 73–74, 199n24, 200nn29–30

Agriculture, British Board of, 114

aircraft, 102–3, 105, 215n107

air service, 100, 103

Aldershot Garrison, 208n76

Alderson, E.A.H., 162–63; *Pink and Scarlet*, 162–63, 237n82

Aldin, Cecil, 225n49

Allenby, Field Marshall Edmund, 219n123; Allenby's Desert Mounted Corps, 119

American Civil War (War Between the States, War of the Rebellion), xi, 84, 90, 98, 101, 110, 115, 197–98n16, 204n48, 205n58, 213n97, 215n104, 222n21

American Expeditionary Force, 123, 226n21

American Indian Wars, xi, 116, 208–9n77

American Revolutionary War, 32, 86

American Volunteer Motor-Ambulance Corps in France, 132

Anderson, Dorothy: "Baker, Valentine [*called* Baker Pasha]," 203–4n47

Angell, Norman: *The Great Illusion,* 238–39n6. *See also* Renoir, Jean: *La Grande Illusion*

Anglesey, Marquess of: *A History of the British Cavalry,* xiv, 77, 202n40, 223n26

"Animals in War Memorial," 127, 224n36, 226n51

arme blanche, 68, 91, 96, 97, 196n10, 206n60, 209n81, 213n97, 217–18n117

arme blanche controversy, xvi, 67, 83, 91, 96–98, 107, 212–13n96, 219n124

armored vehicles, 218n121. *See also* tanks

Army Service Schools (U.S.), 98, 99

Army Veterinary Service (British), 116–17

Astley, John, 14–16, 17, 21, 29, 177n45, 178n53; *The Art of Riding,* 14–16, 18, 176n40, 177n41, 177nn43–44, 177nn47–48, 178n56

Atteridge, A. Hilliard: editor's note to Bernhardi, *Cavalry in Future Wars*

HORSES IN HISTORY

SERIES EDITOR: James C. Nicholson

For thousands of years, humans have utilized horses for transportation, recreation, war, agriculture, and sport. Arguably, no animal has had a greater influence on human history. Horses in History explores this special human-equine relationship, encompassing a broad range of topics, from ancient Chinese polo to modern Thoroughbred racing. From biographies of influential equestrians to studies of horses in literature, television, and film, this series profiles racehorses, warhorses, sport horses, and plow horses in novel and compelling ways.